Abdourahmane Daly Diallo

Geographic information system for a development project

Abdourahmane Daly Diallo

Geographic information system for a development project

Development of a geoportal to improve information dissemination, case of the AgriFARM project, Guinea

ScienciaScripts

Imprint

Any brand names and product names mentioned in this book are subject to trademark, brand or patent protection and are trademarks or registered trademarks of their respective holders. The use of brand names, product names, common names, trade names, product descriptions etc. even without a particular marking in this work is in no way to be construed to mean that such names may be regarded as unrestricted in respect of trademark and brand protection legislation and could thus be used by anyone.

Cover image: www.ingimage.com

This book is a translation from the original published under ISBN 978-620-6-72729-3.

Publisher:
Sciencia Scripts
is a trademark of
Dodo Books Indian Ocean Ltd. and OmniScriptum S.R.L publishing group

120 High Road, East Finchley, London, N2 9ED, United Kingdom
Str. Armeneasca 28/1, office 1, Chisinau MD-2012, Republic of Moldova, Europe
Printed at: see last page
ISBN: 978-620-2-86691-0

DEDICACES

To my late wife Hawa BAH

Called to GOD this November 14, 2020

Dear Hawa, it is with great emotion that I dedicate this memoir to you.

It's the fruit of your support, without which I wouldn't be able to reconcile study, family and professional life. You told me to "go for it! You can count on me".

"Man proposes, GOD disposes, for he is sovereign in his decisions".

"Everything GOD does is good".

Thank you HAWA

REST IN PEACE

AMEN

ACKNOWLEDGEMENTS

My sincere thanks :

- ✓ To Dr TALLA TANKAM Narcisse, Scientific Supervisor ;
- ✓ To Mr. LATE Boniface, Professional Framer;
- ✓ To Mr. WADOUFEY Addel, Professional Coach, for his guidance and coaching efforts;
- ✓ To Pr TCHOTSOUA Michel, Scientific and Pedagogical Referent of the Master GAGER for his leadership throughout this academic year;
- ✓ To the entire GAGER Master's team, for the rigor and quality of the training;
- ✓ To my supervisor Elhadj Tamsir Djibril BANGOURA, Project Coordinator, for his guidance and facilitation;
- ✓ To all colleagues of the AgriFARM project, for their collaboration during the internship;
- ✓ Finally to all the colleagues of the Master GAGER 2019 - 2020 class for the collaboration.

SUMMARY

In Guinea, development projects integrated Geographic Information Systems (GIS) as a planning and decision-making tool in the 2000s. The Agriculture FAmiliale Résilience et Marchés (AgriFARM) project is a Guinean government project, financed jointly by IFAD, OFID and BADEA. It has a GIS linked to the Monitoring and Evaluation System.

The project aims to disseminate information based on the spatial dimension and the Internet for access by the general public. This is why the theme *"Developing a geoportal to improve the dissemination of information, the case of the AgriFARM project"* was chosen.

To achieve this, field and laboratory work is carried out on data acquisition, processing, integration into a database and a cartographic server for publication. A web platform is then developed on the basis of elaborate specifications and user expectations.

Eventually, the entire project planning is mapped and integrated into a GIS database. Relevant information is selected for distribution. The geoportal designed features all the classic functions of interactive cartography. It is accessible to the general public via the Internet at www.geo-agrifarm.com.

KEYWORDS. Geoportal, visibility, AgriFARM project, Upper and Middle Guinea

TABLE OF CONTENTS

GENERAL INTRODUCTION

1.1 Context

In Africa, various issues linked to improving living conditions and sustainable environmental management have been identified at both meso and macro levels. To meet the growing needs of populations, development programs and projects have been set up in various countries. Their mission is in line with the Millennium Development Goals defined by the United Nations.

It is in this context that Guinea, in collaboration with IFAD, OFID and BADEA, has set up a development project which forms part of the sectoral policy drawn up for the country after the health crisis (Ebola). To achieve the project's objectives, a monitoring and evaluation system has been set up, with GIS as one of the planning and decision-making tools.

The Geographic Information System is often used in development projects for static mapping, thus limiting its ability to support communication through the dissemination of information to the general public. Access to the general public enables continuous improvement in the process of disseminating project information to ensure its visibility.

So there's a need for a channel or tool for disseminating localized, large-scale information. GIS, in association with the web, is part of this impetus, through the creation of geoportals. The result of this association is interactive cartography. It is in this context that the theme of this dissertation, entitled "Elaboration of a geoportal to improve the dissemination of information, the case of the AgriFARM project", was initiated.

The aim is to set up a GIS for the AgriFARM project based on open source resources, with a data component organized around a database. This database provides the essential information for publication on the web.

The main objective of this work is to design a geoportal accessible to the general public via the Internet. It allows the publication of project planning information. This is focused on the project's intervention zones, its territorial approach and the key indicators defined at the design stage.

With regard to the dynamic nature of cartography, the platform features functionalities and tools for navigation, search, positioning and surveying on plan and satellite image backgrounds.

To achieve this, we carried out field and laboratory work using mainly open source resources. Next, the data to be published was selected and passed on to the web platform development team. In this context, we worked on the GIS part of the project, while the IT part was entrusted to CDI Guinée for the development and hosting of the geoportal.

1.2 Issues

Geomatics has undergone significant development over the years, enabling universal access to tools such as GIS with the opportunities offered by open source solutions. In the recent past, the use of GIS in development projects in Africa was considered to be costly, with expertise sometimes limited. Today, this barrier is tending to disappear. This has enabled GIS to be increasingly integrated as a planning and decision-making tool, and above all as a means of disseminating project information.

On a macro level, the process of disseminating quality information to the general public using static media seems limited to this area of ICT. The interactive nature of these media remains weak, hence the need to explore new possibilities offering the user considerable leeway when it comes to visualizing and manipulating geographic information. This is where interactive or dynamic cartography comes in.

With the potential for expression offered by Information and Communication Technologies, the rendering of development project websites can be increasingly accessible and attractive. A new use of cartography "allows the citizen to be involved in the making of the map", gives the possibility "to display layers and their graphic representation according to scale" and "interactive links between the map and multimedia documents can be created". They also enable "more realistic cartography with 3D and/or animation". E. CHESNEAU and J. ULTSCH, (2010)

In the Guinean context, the use of geomatics has been impacted by its late introduction into the professional and academic domains. For many years, geomatics was practiced by a limited number of users, generally trained on the job (mastering only GPS data collection and the use of GIS software such as Mapinfo for a long period, then Arcgis, etc.). This deficit led to the following observations:

- In general, the use of GIS in development projects is limited to static cartography, which in turn limits the number of (mainly external) users. Expanding the number of users responds to the need to disseminate information regardless of distance and geographical location;
- Failure to take GIS into account in the design of certain projects (due to a lack of awareness of the potential of geomatics tools);

- Poor integration of GIS in the process of capitalizing on project achievements, resulting in the overlapping of project intervention zones in the field (impact on the implementation of rural development policies).

1.3 Scientific background

The development of NICTs and the ease of access to resources (tools and materials) have made it possible to set up a dynamic mapping service via the Internet. This has led to research and applications in a wide range of fields, from education to communication and large-scale information dissemination. Several researchers have taken an interest in WebSIG.

Indeed, J. P. MBAHA and G. B. TCHOUNGA (2018) explain the challenges of using online and offline interactive mapping for the benefit of natural risk management on the Cameroonian coast. They present the difficulties of access to information by populations in developing countries to justify the need for the use of wepmapping.

On the same subject of dynamic cartography, I. BALDE (2008) discusses the use of open source resources to implement interactive cartography. The aim of this work is to enable maps to be displayed at the user's request (man-machine interaction). It also aims to position the strategic zones of the rural world on it, knowing their geographical coordinates. In this study, these strategic zones are essentially composed of basic infrastructures (education, health, access to water and administration). They are supplemented by a representation of the cadastre of certain rural areas, divided into parcels.

In terms of functionality, the application developed by I. BALDE (2008) also provides the user with map navigation tools. These tools include zooming, moving, displaying area information and displaying the map legend. This project was initiated in Senegal as part of a program to support local governance based on effective communication. It provides communities with widespread access to information and communication technologies (ICT) to boost their economic and social development.

From setting up websig platforms dedicated to rural development projects, Action Contre la Faim (2017) has been working along these lines in Niger. A webSIG platform has been set up to help with planning, disseminating information and capitalizing on the project's achievements. Called Geosahel.info, this platform is an interactive tool that allows data to be manipulated against a map background, according to the needs and choices of users.

Also in the field of rural development, O. DUPARS-TESSIER (2016) has carried out a study to develop websig in natural resource management, focusing on the development of non-timber forest products. His contribution is primarily aimed at stakeholders involved in the sustainable development of NTFPs. It proposes a platform that offers the possibility of visualizing and manipulating data specialized to the case of NTFPs. It offers tools for locating and validating high-potential sites. This portal introduces editing, analysis and spatial modeling applications exclusive to the NTFP theme. It houses interactive and interoperable data layers, providing easy access to information.

On a professional level in Guinea, we were also interested in the use of GIS in rural development projects (agriculture, infrastructure). We found that the PNAAFA project, run by the same donor (IFAD), had a GIS and a website. However, the GIS was only used for static mapping of the project's interventions. In the age of NICTs, this hinders the dissemination of information, which is the main way of making project achievements visible.

Like this project, others have websites with content based on text, photo images or, in some cases, maps in image format. The interactive nature of the tool is not sufficiently exploited, and it is through this that the user finds himself involved in the choice of information he wishes to visualize. From a scientific and professional point of view, in view of the various contributions and works on the subject of setting up GIS portals, this research project will be applied to the AgriFARM project, integrating a GIS into the monitoring and evaluation system in order to be able to visualize the results on cartographic media.

It deals with the setting up of a geoportal for a rural development project based on agriculture and infrastructure in the context of the Republic of Guinea.

It contributes to improving the practice of GIT in rural development projects through interactivity in the dissemination of information. It also aims to promote open source resources and the integration of dynamic cartography into rural development project websites. In the long term, it will facilitate the capitalization of AgriFARM project actions on IFAD's geonode platform (geoportal of IFAD projects across Africa and the world).

1.4 Research questions

Main question: How can geographic information technologies help to improve the dissemination of information to enhance the visibility of the AgriFARM Upper and Middle Guinea project?

Q1 - Can geomatics tools be used to map the activities of a development project?

Q2 - Does a geoportal platform ensure the dissemination of information on development project interventions?

Q3 - Does the availability of geographic information on the Internet contribute to the visibility of a project's actions?

1.5 Objectives

The overall aim of this research theme is to improve the information process relating to AgriFARM project activities, by setting up an interactive cartography via the Internet.

To achieve this goal, specific objectives have been set:

- ensure the geolocation/georeferencing of activities planned by the project throughout its intervention zone (Upper and Middle Guinea);
- process and integrate data into a GIS database and map server for publication;
- make geographic data accessible on a web platform designed for this purpose, with classic dynamic mapping functionalities such as visualization, manipulation, information search and connection to online resources.

1.6 Research hypotheses

The main hypothesis is that geographic information technologies and the Internet, through a geoportal, can be used to disseminate information to make development project activities more visible.

Specifically, three secondary hypotheses emerge.

H1: geomatics tools can be used to geolocate, process, analyze and disseminate all information linked to project activities.

H2: a GIS database provides the information required for publication

H3: Web programming tools can be used to design and distribute geographic information generated by GIS.

1.7 Plan

This document presents the work accomplished during the six (6) months of internship. It is structured as follows. The general introduction (above) includes the problematic of the subject, the context, the scientific questioning, the objectives and hypotheses.

The second part deals with methodology, focusing on field and laboratory equipment and methods. It describes the entire process, from GIS data acquisition to the development of the web platform.

Finally, the last part presents the result, i.e. the cartographic data, the GIS database and the geoportal with all its functionalities. The cartography presented as a result concerns the national territory of Guinea and the interventions of the AgriFARM project. And the geoportal is presented here from the point of view of structure and operation, i.e. the rendering that the Internet user will see in front of him on a browser.

The document also includes an appendix containing all relevant data referenced in the chapters. Important data such as the MCD and the geoportal source code are included for information purposes.

To sum up, the rest of this document is structured into three chapters and a general conclusion.

Chapter I: Background and study framework
Chapter II: Materials and methods
Chapter III: Results, interpretation and discussion
General conclusion.

CONTEXT AND FRAMEWORK OF THE STUDY

This chapter introduces the context for an understanding of the study framework. It deals with the study area through the AgriFARM project and its intervention zone. It then presents the scientific background to the process of developing a geoportal for information dissemination, and all the key concepts associated with it. In this section, an example of a geoportal is presented for illustrative purposes.

In the geographical and social context, the AgriFARM project under study is extensively presented in relation to Guinea. Spatially speaking, the areas and logic of intervention are also explained in detail.

1.1 Geographical and social context

1.1.1 Location and setting

The "***Agriculture FAmiliale, Résilience et Marchés en Haute et Moyenne Guinée (AgriFARM)***" project is a Guinean project under the supervision of the Ministry of Agriculture of the Republic of Guinea. It benefits from financing agreements between Guinea, IFAD, OFID and BADEA for a total cost of 120.2 million US dollars.

The overall objective of the project is to contribute to a sustainable improvement in the food security and nutritional situation and crisis resilience of rural households in the regions of Upper and Middle Guinea.

The project's development objective is to sustainably increase the incomes of 65,000 family farms, their resilience to external shocks, including climate change, and improve their nutritional status, as well as their access to local, urban and regional markets in the 15 targeted prefectures of the Upper and Middle Guinea regions.

The AgriFARM project is organized into two complementary components: the first component, "Strengthening family farming and resilience to climate change", is broken down into four sub-components whose activities place family farming at the heart of interventions, through sustainable development of sub-watersheds, structuring water management schemes in production basins, improving agricultural productivity, strengthening grassroots rural organizations and associations, and improving the nutritional situation.

The second component, "Market Access", comprises three sub-components whose activities aim to ensure outlets for surplus production, by building/rehabilitating semi-wholesale markets, collection markets and rural tracks, by setting up management systems for these economic infrastructures to ensure their sustainability, and by supporting the financing of agricultural and rural entrepreneurship. A third component will cover project management and coordination, monitoring and evaluation, and knowledge management.

A third component covers Management, Coordination, Monitoring-Evaluation and Knowledge Management. It is at the level of this component that the said internship for this research dissertation takes place. *Source : DCP AgriFARM 2018*

Organizational framework

The Ministry of Agriculture (MA) oversees the AgriFARM Project. The Joint Steering Committee for IFAD co-financed projects (AgriFARM, PNAAFA-BGF) is appointed with a maximum of fifteen members, in accordance with the Government's recommendations.

Project coordination and management is entrusted to a Project Coordination and Management Unit (UGP), which is autonomous in programming, budgeting and financial management under the authority of the Ministry of Agriculture. The head office is based in Mamou, with two branches in Labé and Kankan. The head office is also directly responsible for activities in the Mamou region. In the Boké and Faranah regions, where the PNAAFA-BGF project is active until 2019, particular attention is being paid to the synergies to be sought with the AgriFARM project.

Implementation is based on "faire-faire" and on strengthening the specialized agencies and decentralized technical services of the Ministry of Agriculture and its partners, through agreements developed on the basis of results-based management. International technical assistance is available in key areas of intervention during the first years of implementation (management, fiduciary, monitoring-evaluation, watershed management, social engineering, civil engineering). *Source : DCP AgriFARM 2018*

1.1.2 Geographical location

The Republic of Guinea is a coastal country located in the western part of the African continent, midway between the Equator and the Tropic of Cancer (7° 30' and 12° 30' North latitude and 8° and 15° West longitude). Covering an area of 245,857 km2, it is bordered to the west by Guinea Bissau and the Atlantic Ocean, to the north by Senegal and Mali, to the east by Côte d'Ivoire and to the south by Sierra Leone and Liberia.

From a geo-ecological point of view, Guinea is subdivided into four fairly distinct and internally homogeneous natural regions. The country owes this originality to its natural environment, which is characterized by climatic contrasts, mountain barriers and the orientation of relief, all of which combine to give each region its own particularities in terms of climate, soils, vegetation and way of life.

Moyenne Guinée or Fouta Djallon is a region of mountains and plateaus. Its highest peak is Mount Loura (Mali) at 1,538 m. The massif is deeply incised by valleys with interior plains and depressions. Heavily degraded soils are gradually being replaced by bowé, narrowing the expanse of farmland.

The region's numerous rivers and streams make it the water tower of West Africa. However, these rivers are enclosed in deep valleys, which means that the plains on which they flow are narrow, making hydro-agricultural development difficult.

The tropical climate is modified into a mountain microclimate. Rainfall is very low. This is an area of grazing, citrus groves and vegetable gardens. Today, the degradation of the environment has led herders to extend transhumance to Lower Guinea (Boké, Boffa and Forécariah), whereas it was originally practiced between the high plateaus in the rainy season and the depressions in the dry season. Because of its mountainous terrain and the extent of its ecosystem degradation, Middle Guinea is the poorest region from an agricultural point of view.

Upper Guinea is part of a morphologically and climatically extensive geographical unit. It is a region of savannahs and plateaus. The Niger River and its tributaries have carved out wet, terraced plains that are ideal for flooded rice cultivation.

In terms of climate, this is the driest region of Guinea. Rainfall varies between 1,200 and 1,800 mm per year. The dry season is longer (7 to 8 months) and average temperatures are relatively high for most of the year. Maximum temperatures sometimes exceed 40 ° C in March-April. The vegetation is punctuated by thin forest galleries.

Despite the existence of vast rice-growing river plains, Upper Guinea is not a prosperous agricultural area, due to frequent droughts and the infestation of certain river valleys by similis, vectors of onchocerciasis (Tinkisso valley). On the other hand, it is a favored area for river fishing and livestock breeding. Artisanal gold and diamond mining is a traditional activity for the people of this region. *Source : DCP AgriFARM 2018*

The map below shows the study area (Upper and Middle Guinea) within Guinea.

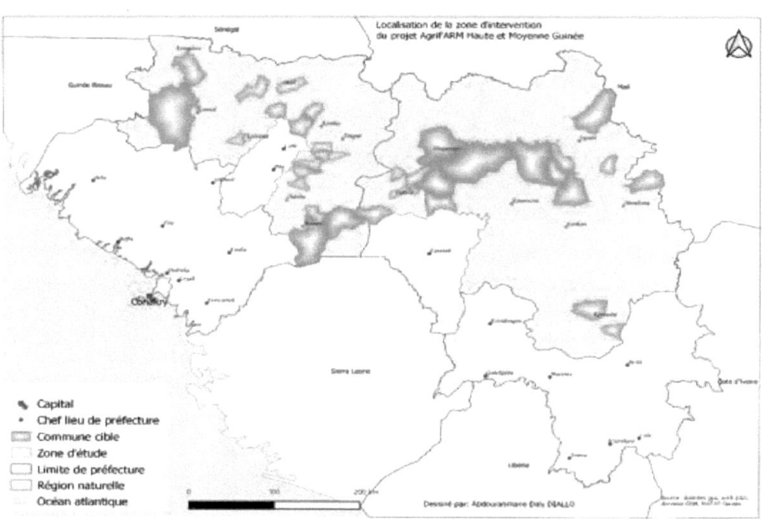

Figure 1: Presentation of the study area

1.1.3 Socio-economic situation

According to the AgriFARM 2018 Project Concept Paper (PCP), Guinea had a population of around 12.7 million in 2017. The country's largest employer, the agricultural sector plays a key role in poverty reduction and rural development: it provides income for 57% of rural households and employment for 52% of the workforce. World Bank, (2017)

- **Growth stalled in 2019 at 5.6%**, after mining and non-mining activities slowed due to moderate growth in the agricultural sector. *Source: DCP AgriFARM 2018*

- Over the same year, **inflation is estimated at 9.5%**, down slightly on 2018 (9.8%), due to a tightening of monetary policy by the Central Bank. *Source: DCP AgriFARM 2018*

- The **overall budget deficit (including grants) improved** from 1.1% of GDP in 2018 to 0.9% in 2019. Tax revenues, which were 0.6 percentage points lower than forecast in the Loi de Finances Rectificative, barely increased to 12.5% of GDP in 2019. Mining tax revenues fell from 2.6%

to 1.9% of GDP between 2018 and 2019 due to lower production mining activities in the second half of 2019. *Source: DCP AgriFARM 2018*

- Over the same period, public subsidies for electricity rose from 0.8% to 1.7% of GDP, due to payment for electricity supplied by the Kaleta Dam. At the same time, the government cut capital expenditure, which was 3.1 points of GDP lower than budgeted, in order to limit the budget deficit. *Source: DCP AgriFARM 2018*

- Public debt fell from 37.3% to GDP in 2018 to 34.4% in 2019, due to the delayed implementation of externally financed projects. The risk of public debt overhang remains moderate. *Source: DCP AgriFARM 2018*

- The COVID-19 epidemic is exposing the country to health and economic shocks that will impact on household well-being. The livelihoods of vulnerable people, who often work in the informal sector and in small and medium-sized enterprises, are likely to be particularly compromised. Especially as these people have limited savings and access to financial services to cope with the crisis. *Source : DCP AgriFARM 2018*

1.1.4 National context, development and poverty in rural areas

The AgriFARM project was born out of Guinea's post-Ebola socio-economic and political context.

a. Overview of the Guinean economy

Ranked among the least developed countries with low incomes, Guinea suffered the full force of the Ebola crisis (February 2014-June 1, 2016). With over 2,500 deaths, the epidemic impacted the country's economy creating, among other things, food shortages, significant price variations4 and factory shutdowns. Growth only resumed in 2016, driven by the mining industry and supportive public policies. The mining industry stands out as the country's economic engine, accounting for up to 90% of exports (gold, bauxite). However, job creation in this sector remains limited (less than 2.5% of the working population), with consequent socio-environmental risks. *Source : DCP AgriFARM 2018*

b. Public policy.

In this post-crisis context, Guinea is in the process of updating its public policies. The document "Guinée Vision 2040" was adopted in April 2017 and outlines the country's development directions, along with the National Economic and Social Development Plan (PNDES) 2016-2020 and the National Agricultural Development Policy (PNDA).

The provisional version of the NADP identifies three challenges: i) increased access to buoyant markets, particularly those in the sub-region and on the continent; ii) high productivity; iii) effective governance of the agricultural sector. The agricultural sector is expected to increase its contribution to the national economy by 50%. The government is committed to allocating 12.5% of its budget to agriculture, up from the current 7.3%. *Source : DCP AgriFARM 2018*

c. Agricultural potential and urbanization

Guinea has considerable agricultural potential, which remains under-exploited. The soil and climate conditions are highly conducive to agriculture, and only 25% of its arable land potential (6.2 million hectares) is exploited. Despite this potential, the country's cereal yields are low, at 1.25 t/ha compared with an average of 1.45 t/ha in West Africa.

Guinean agriculture is predominantly family-based, and remains concentrated on food crops, notably cereals (rice and corn), tubers and palm oil. Rice is the leading national production, with around 2 million tonnes of paddy in 2014/2015. The agricultural sector grew at an annual rate of 5% between 2011 and 2013, accounting for 20% of Guinean GDP, driven mainly by the increase in cultivated land (+10% per year)10.

Growing demand for food products is supported by demographic growth (2.5% in 2016) and urbanization (38% of the population in 2016 versus 33% in 2006). 4. Agricultural markets and processing. Agricultural produce is rarely marketed in suitable areas, notably due to a lack of collection and processing infrastructure. Agricultural produce suffers high post-harvest losses, of 22% for rice and 20% for maize, while high transaction costs, linked to transport difficulties, have a negative impact on field purchase prices, particularly during the rainy season.

Insufficient storage capacity, combined with production shortfalls and producers' immediate financial needs, often force producers to sell their produce at harvest time, when prices are at their lowest, which does not encourage them to increase their production volume the following season. As a result, volumes traded, particularly for food crops, are unable to meet growing demand.

In 2014, rice imports, equivalent to 11% of total imports, accounted for 20% of consumption, although Guinean consumers have a preference for local parboiled rice, which represents an opportunity for Guinean rice. The maize market is booming with the development of poultry farming. Processing is carried out almost exclusively in small-scale units, which nevertheless handle only a quarter of agricultural production and represent an opportunity for this sector.

Nutritional situation. Nearly a third of Guinean children under the age of five suffer from chronic malnutrition (WHO 2015), with a stunting rate at birth of 9% (EDS 2013). One of the major factors behind these high thresholds is explained by poorly varied diets. This low diversity also leads to inadequate nutritional states in both mother and child11. The double burden of malnutrition is also very present, with 13% of women having Body Mass Indices (BMIs) below 18.5 and 14% of women with BMIs above 25 (WHO 2015), leading to an upsurge in cardiovascular disease and type 2 diabetes (8% in rural areas).

The low diversity in household food choices is due to shortcomings concerning: (i) the use of natural resources and the availability of commodities such as vegetables, fruit, legumes and animal proteins, particularly during the lean season (EDS 2013); (ii) the preservation and processing of perishable products, pushing households to sell fresh (Guinea Nutrition Assessment, USAID 2015); (iii) income levels not allowing diversification of the food basket, and difficult physical access with isolated and poorly supplied markets; (iv) knowledge of meal preparation, consumption habits, household meal distribution and gender aspects; and (v) care and feeding practices for children under 5, including the non-existence of complementary foods. *Source: DCP AgriFARM 2018*

d. IFAD country programme overview

IFAD has invested some US$200 million in 13 projects since it began operations in Guinea in 1980. The only current project, the Programme national d'appui aux acteurs des filières agricoles (PNAAFA extension Basse-Guinée et Faranah), with total funding of US$23 million, will come to an end on December 31, 201913. With project management delegated to the OPAs, the PNAAFA BGF is the last phase of the national PNAAFA program, which began in 2010.

The last RB-COSOP ended on December 31, 2014, and its renewal, at that time, could not be achieved, given the Ebola-related health crisis. In early 2017, the Government and IFAD nevertheless adopted a Country Strategy Note for the 2017/18 period to align with the national public policy review timetable.

In this context, the Government of Guinea has requested IFAD's assistance in financing a new investment program in the agricultural sector, estimated at US$ 100 million, to support the development of family farming and the implementation of national policies and strategies under the National Plan for Agricultural Investment and Food and Nutritional Security (PNIASA), itself aligned with the National Economic and Social Development Plan14 .

In response to this request, IFAD proceeded jointly with the Government of Guinea to formulate the new Project entitled Projet pour l'Agriculture Familiale, Résilience, Marchés (AgriFARM). After two joint missions, from November 1 to 24, 2017 and from January 14 to 31, 2018, the Government of Guinea and IFAD defined the intervention area in 15 prefectures of the Upper and Middle Guinea regions. *Source : DCP AgriFARM 2018*

e. Intervention logic.

The Project aims to enhance the contribution of family farming to inclusive national economic development, while ensuring household food and nutritional security and resilience to climate change. Based on the development of dynamic family farming, the Project will support an economic model that is competitive and market-oriented (supply/demand, infrastructure, public-private partnerships), inclusive (women, young people, OPAs), sustainable (resilience to climate change) and guarantees food security and improved nutritional status.

By capitalizing on growing domestic demand, this strategy will have the dual function of (i) lifting the target population out of poverty, ensuring their food security and improving their nutritional situation - and, (ii) contributing to the country's economic development objectives.

The Project's approach is based on three intervention principles: i) improving food security and nutrition, taking into account the dimensions of availability, access and use of food through sustainable systems; ii) systemic, demand-driven; iii) territorial continuity of interventions, organized around a coherent entity following the flow of agricultural products.

Improving household food security and nutrition through intensive, market-oriented family farming. This will be achieved by: i) increasing the income of family farms, whether through higher volumes marketed or higher margins; ii) increasing the availability of food (market access, greater auto-consumable production) for the beneficiary and the country; iii) increasing the quality of food (more diversified, controlled sanitary practices) for the beneficiary and the country.

By seeking to address the four pillars of food and nutritional security - availability, access, use and stability of food - the Project adopts an approach that combines improving the productivity of family farming production systems with improving the nutritional situation of households. The technical interventions consider rice and maize as priority commodities, as well as leguminous and market garden crops, combined with small-scale livestock farming and pisci-riziculture, and will improve the availability of agricultural products on farms.

Increased income from easier access to the market will also facilitate access to food purchases to complement those produced and consumed by the household. Improved utilization by both adults and children will be the focus of awareness campaigns based on food practices that can be improved. Finally, through all its components guaranteeing the sustainability of the systems and transformations, the Project will contribute to an improvement in the nutritional situation.

A systemic approach comprising: i) increasing demand by supporting the disposal of surplus agricultural products from family farming; and ii) increasing supply by supporting increased production. By improving the economic infrastructure - tracks, collection markets and half-wholesale markets - as well as social and technological capital, the Project aims to control transaction costs and increase the volume of cereal trade from family farming to consumption centers (towns, mining basins). It will ultimately aim to: i) make trade more fluid and increase it through more efficient value chains driven by dynamic demand; ii) promote trade on a fair basis with standardized measurements, a multiplicity of offers and the availability of information; iii) support the formalization of market players, to ultimately provide them with new market opportunities (broader, more direct, etc.).

A territorial whole. This approach is based on taking into account the territorial space targeting family farming and all the players in the main food commodity chains (producers, brokers, traders, transporters, processors, etc.). This territorial unit of economic development is understood as an area where economic activities linked to the production, processing and marketing of the main agricultural products from adjacent production basins are concentrated. These territorial units of economic development are made up of production basins (and their catchment areas), strategic rural routes for the sale of products, and semi-wholesale markets linking them to consumption centers.

This approach will also make it possible to concentrate investments: i) geographically, by targeting production basins, rural tracks and semi-wholesale markets; ii) strategically, by synchronizing support for production and its sale, so that synergies between demand and supply can be expressed (see above). The diagram below summarizes this approach. *Source : DCP AgriFARM 2018*

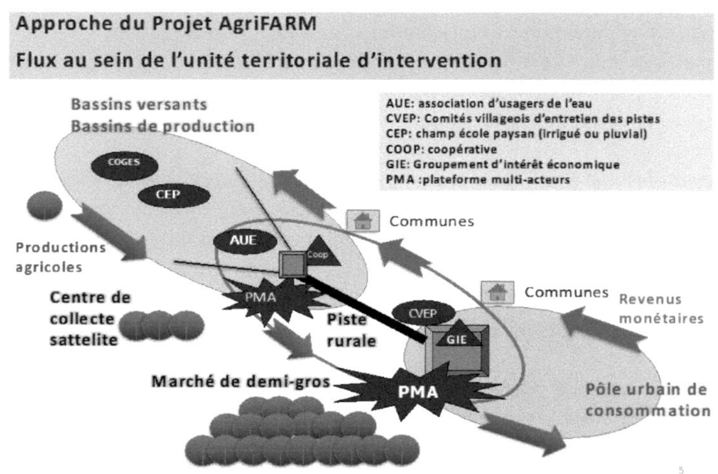

Source: DCP AgriFARM 2018

Figure 2: AgriFARM project approach

1.2 Key concepts

1.2.1 Geographic Information System (GIS)

1.2.1.1 Definitions

GIS is made up of three words, each with its own meaning. As a tool based on the representation of existing on the earth's surface, GIS is used across the board. As a direct consequence, different definitions of GIS are used in different fields of application and different schools of thought. Thus, for this project, the following definition is taken as a reference.

GIS is "an organized set of hardware, software, geographic data and personnel capable of capturing, storing, updating, manipulating, analyzing and presenting all forms of geographically referenced information" De Blomac et al. (1994) quoted by J.M. FOTSING, (2005).

From the diversity of definitions, questioning also helps to understand what GIS represents. Answers to the questions Where? What? How? When? What if? help define GIS.

According to M. TCHOTSOUA (2019), GIS definitions are all based on the three words that form its acronym GIS namely:

- System" refers to a collection of elements satisfying a relationship ;

- Information" is the potential meaning attached to data which is itself a conventional representation of a fact, notion or phenomenon suitable for communication, interpretation or processing;

- "Geographic" refers to the land, people and their activities on the earth's surface.

From these three words, information and geography (IG) have been associated to designate "data with a spatial reference either in the form of : 1. geographical coordinates 2. Place name 3. Postal or other address", Y. KOUBA (2018). He adds, GI can be duplicated without deterioration, exchanged at high speed through web networks, cumulated with different information with a view to producing new information.

On this subject, economist M. DIDIER (1990), in a study commissioned by the Comité National de l'Informatique Géographique (CNIG), gives a definition of geographic information that specifies its objectives. For the author, it is a matter of making available a: "Set of spatially referenced data, structured in such a way as to be able to conveniently extract summaries useful for decision-making".

According to H. CHAKROUN (2014), the need to set up instruments to monitor and manage our territory and geographical environment appears every day to be an inescapable task. As R. Caloz (1997) "to manage is to predict, and predicting the evolution of a phenomenon presupposes knowledge of its dynamics, and therefore of the parameters that determine it and their respective roles." This management requires operations representing a considerable volume of data on the territory and processing capacities of the same order. This is what gave rise to GIS (Geographic Information Systems).

Over the past five decades, GIS has evolved from a concept to a science. The phenomenal evolution of GIS, from a rudimentary tool to a modern and powerful platform for understanding and planning our world, is marked by several key milestones.

1.2.1.3 Components

From the above definitions, GIS is presented as a system, i.e. with segments or components. It is made up of five components:

> **Hardware**

This component of a GIS is made up of several elements whose scope of action extends from data collection (GPS), processing (computers), storage (servers) to information dissemination. These different elements evolve with technology, making GIS tasks easier and easier.

> **Software**

To operate the hardware, the GIS needs software. Depending on the status of the license, this may be paid for or free. In terms of functionality, all GIS software has similar tools for functions ranging from acquisition to real-world representation or abstraction. It is through the use of software that the five A+1s are applied, namely: Acquisition, Archiving, Analysis, Display, Abstraction. The prospecting or anticipation function accompanies the five main functions.

In this era of open source solutions, some software such as Quantum GIS is being exploited. It's free and distributed under the GNU GPL (General Public License). This allows users to study and modify the source code, with guaranteed access to a GIS program. It is available in several versions (from 1 to 3.14 in July 2020). It is an official project of the Open Source Geospatial Foundation (OSGeo). OSGeo is a non-profit association founded on February 27, 2006 in Delaware (USA).

For the purposes of this thesis, QGIS is used as the GIS software. It is available as a free download from the official website. It is easy to use and offers a wide range of functions, as well as access to complementary extensions. It is also compatible with all operating systems (Windows, Linux, etc.), servers and database systems (RDBMS such as posgresql/postgis).

> **Data**

In the presence of hardware and software, data is an essential element in ensuring system operation. As such, it can be likened to fuel for the vehicle.

For a GIS, there are two categories of data: rasters and vectors. In practical terms, rasters are images obtained by digitization and taken using devices or equipment placed at a certain height above the ground.

According to A. DENIS (2020), there are two main types of geographic data: matrices and vectors.

Matrices", more commonly known as "rasters", are "grids made up of cells", or in other words, "images made up of pixels". Each cell contains a value, often representing a geographical phenomenon such as altitude or land use. Examples include scanned maps, aerial photographs, satellite images, digital photos, DTMs (Digital Terrain Models), etc.

Vectors" are made up of points, lines or polygons. The most common vector file formats used in GIS are "shapefile" and (more recently) "GeoPackage". The term "shapefile" is often used as a synonym for "vector file".

Vector data represents the translation of field observations into geometric shapes, which can be point, linear or surface.

- Point objects: a point is represented geographically by its longitude (x) and latitude (y). For example, the capital of a prefecture can be represented by a point.
- Linear object: defined by a succession of points with geographic coordinates (x1, y1), (xn, yn). Roads and courtyards can be represented by lines.
- Surface object: defined by a succession of closed points (superimposition of start and end points) with geographic coordinates (x1, y1), (xn, yn). An area is represented by a polygon (e.g. a plain, a pond).

> **Users**

A Geographic Information System is a tool that is operated by a community of users. This community ranges from those who create and maintain systems to users of geographic information. The number of users in this community is growing all the time, thanks to the rapid expansion of NICTs.

> **Methods**

The implementation and operation of a GIS is based on hardware, data, users and software. Given the diversity of these components, the operation of a GIS follows rules, procedures and standards that may vary from one organization to another.

1.2.1.4 Functions

"GIS technology uses geographic science with tools for understanding and collaboration. It helps users achieve a common goal: obtaining actionable information from all types of data." Esri France (2020)

This makes GIS a dual tool for management and decision-making. A GIS has a cross-cutting (different domains) and horizontal aspect when it comes to the actors involved in the management and use of space, i.e. from technicians to decision-makers.

As mentioned above in the software section of GIS components, GIS functionality can be summed up in 5 A's + 1. Some authors use these 5 A's to characterize a GIS.

It is in this context that M. TCHOTSOUA states: "Although a definition that is unanimously accepted by all is still difficult to find, we are beginning to note a convergence among authors on a certain number of points that FOTSING (2005) summarizes as follows. "*A GIS must be able to answer the five basic recurring questions:*

- *Where: where is this object, this phenomenon?*

- *What: what can be found there?*

- *How: what relationships exist between these objects or phenomena?*

- *When: when did changes occur?*

- *If: what would happen if such and such an evolutionary scenario were to occur?"*

On this basis, we can say that a geographic information system is: "a system capable of acquiring, archiving, providing access to, analyzing and displaying geographically localized information". This brings out the 5A+1 functionalities, which are: Acquisition, Assembly, Archiving, Analysis and Display + Abstraction.

Thus I. BALDE (2008) describes the functionalities of a GIS in the following terms.

Abstraction: to represent the real world, data is organized into graphical objects or geometric elements characterized by attributes. These objects are linked together to interpret the real world, i.e. themselves and their environment.

Acquisition: this **is** the process by which the GIS is supplied with vector or raster data. Here, data is made available in digital format.

Archiving: this is the management of data in a database and its mobility from a workspace to storage.

Analysis: this refers to the manipulation and interrogation of data for query purposes (spatial, attribute, etc.).

Display: Its purpose is to enable the user to grasp spatial phenomena, as long as the graphic representation respects the rules of cartography.

a. How it works

GIS works on the principle of superimposing geographic layers. This principle is embodied in each software project, for example qgis (QGIS) or mxd (ARCGIS) or wor document (MapInfo).

1.2.2 Notion of geolocation

Geolocation is the process by which an object or phenomenon is located on a map using its geographic coordinates. In addition to longitude and latitude, a third dimension has come into play with the measurement of altitude. All this is made possible by the Global Positioning System (GPS). The evolution of technology has democratized the notion of geolocation with a transversal dimension. The example of Android is a case in point. Satellites, Wi-Fi, GSM, Bluetooth signals and IP addresses are all geolocation techniques.

1.2.3 GPS

According to the Office québécois de la langue française (2008), GPS is a "Positioning system that makes it possible, at a precise moment, to determine the geographic position of an object or a person by using signals emitted by satellites, placed in orbit around the Earth, towards a receiving device located at the place to be located."

Originally, gps was designed by NAVSTAR (*Navigation System using Time and Ranging*) for military use (US army). Over time, its use was extended to civilians. Other systems are currently being developed worldwide, such as *Galileo* in Europe and *Beidou* in China. *Glonass* is another satellite positioning system developed by Russia.

As a system, GPS consists of three segments:
- Space segment
- Control segment
- User segment.

1.2.4 Geomatics

According to the Office québécois de la langue française (2004), geomatics is a "discipline whose object is the management of geographic data and which calls upon the sciences and technologies related to their acquisition, storage, processing and dissemination".

Geomatics uses computer technologies to process and represent geographic information. Its aim is to represent all objects on the earth's surface, as well as the events that occur there. With the evolution of technology, it is increasingly playing other essential roles in understanding and planning territories. These include the projection of actions within a territory, as well as impact measurement and monitoring.

1.2.5 Geographic information (GI)

Geographic information refers to a set of data that can be located in a given area. It can be represented in the form of maps, tables and images. The representation of information in map form is linked to graphic elements (points, lines and polygons).

For this study, geographic information covers :
- the territory defined by the administrative division of Guinea;
- the project area ;
- the location of planned activities;
- road and market infrastructures in the project area.

Geographical information is obtained by analyzing and interpreting certain data. In the case of satellite imagery, following classification, information is collected and integrated into a database. It is then represented on a map that can be used by a wide range of users.

1.2.6 Cartography

Cartography is the set of techniques and operations used to produce maps. To this end, it uses data or results from direct observation or documentation. According to the encyclopedia, a map is a simplified, conventional, flat geometric representation of all or part of the earth's surface, in a suitable similarity ratio, known as scale.

Following the evolution of the discipline over time, the International Cartographic Association defines cartography "as the discipline concerned with the design, production, dissemination and study of maps". It has evolved from static to interactive or dynamic cartography.

1.2.6.2 History of cartography

From the history of cartography, it's worth recalling some of the sciences associated with it. These are: geodesy, surveying, photogrammetry, remote sensing, geographic information systems (GIS), global positioning systems (GPS), mathematics and statistics. Over the course of time, mapping techniques have undergone major changes, with the gradual transition from manual production to automation.

Over the centuries, cartographic production has been constantly influenced by technological change. The old techniques of pen and ink have been replaced over time by computer and software. As a result, the quality of cartographic rendering has improved, saving time and production costs.

In recent years, the development of multimedia and the introduction of virtual reality have joined the cartographic experience. These disciplines are all separate, have common characteristics and share one important aspect, which is their relationship with cartography. This has motivated the development of cartographic components in certain disciplines, a basic knowledge of which forms an important part of modern cartography.

Today, cartography is characterized by two elements. Firstly, its cross-disciplinary use makes this discipline a pillar of our civilization, as it is difficult to imagine the practicability of any field of activity on the earth's surface without a map. Faced with the diversity of challenges facing humanity, particularly the environment, cartography has never ceased to provide solutions, especially when combined with other tools (GIS, remote sensing).

Secondly, the dynamic nature of cartography offers a second essential characteristic. The discipline of cartography is at the heart of a revolution spawned by advances in computer technology.

1.2.7 Interactive mapping

1.2.7.1 Definitions

According to P. MARTIN (2004), "the development of high-performance cartographic systems and the rapid expansion of means of communication linked to various information technologies have opened up new horizons, such as cartographic distribution on the Web, which seems destined for a great future. Indeed, while the popularity of GIS software can be explained by a range of products in line with users' needs, the forecast market share of GIS integrating Web technologies, such as the prototype developed in this thesis, is growing steadily, and should reach 67% of all GIS on the market in 2004, compared with just 15% in 1999".

Since 2005, many leading Internet companies such as Yahoo, Microsoft and Google have been developing online mapping services. They are accompanied by APIs that enable developers to integrate background maps from their source into their online GIS applications. These Internet giants have been joined in their initiative by state bodies such as IGN in France and the Ordnance Survey in the UK.

Today, the services offered by these organizations are growing in number. They allow developers to add their own data for representation on served map backgrounds. From a functional point of view, interactive tools (zoom, click reaction, distance calculation, classification of spatial phenomena, etc.) with the map are made available. R. P. TATSO (2011)

According to D. GADIOU (2012) asserts, "GIS, historically dedicated to experts, are moving towards the general public in various more or less simplified forms. When associated with Web 2.0 practices, Internet users can interact, i.e. read and write maps, which tends towards a new concept of participatory GIS."

The following terms are associated with consumer GIS, so it's worth defining them.
-   Neo geography (new geography) is a trend characterized by the practice of creating personalized online maps from a variety of data sources and tools. It's a new practice made possible by the development of online mapping tools such as Google Maps and Open Street Map.

- The geoweb is a general term for any application or service related to geographic location.
- WebMapping describes dynamic online applications that result in a cartographic representation. WebMapping is a special kind of geoweb application, and is of particular relevance to neogeography.

On the emergence of interactive cartography, G. AYEL (2013) states
"The paper form has its limits, and users are looking for more interactivity to communicate and disseminate information in a more entertaining way, based on interactive cartography. An interactive mapping interface must be simple and intuitive. Interactive tools and design must be based on classic webmapping applications.

1.2.8 Online mapping concept

According to I. BALDE (2008), online cartography meets real needs for rapid dissemination of information and remote updating of data. Although the cartographic result facilitates understanding of the surrounding space, the implementation of such platforms requires cross-disciplinary skills in both computing and geography.

Online cartography, developed via the Internet, is a form of digital cartography. It uses the Internet to produce, design and publish geographic information through maps. This has led to the development of several mapping services, such as google maps, google earth and bings maps. The Open Street Map service is a free cooperative solution.

1.2.9 Internet

1.2.9.1 Definition

According to Wikipedia, the Internet is the global computer network accessible to the public. It is a packet-switched, hubless network of networks, made up of millions of public, private, academic, commercial and government networks, themselves grouped into autonomous networks (there were over 91,000 in 2019).

The Internet was popularized in the 1990s with the emergence of the World Wide Web. This led to the emergence and growth of connected objects (2010). Information is transmitted via the Internet using a standardized set of data transfer protocols, enabling a variety of applications such as e-mail, the World Wide Web, instant messaging, peer-to-peer, streaming and teleconferencing.

The Internet is attracting an ever-growing number of users. Access to the Internet is supported by a provider via various means of electronic communication. These can be wired or wireless, such as 4G or 5G.

1.2.9.2 Web 2.0

According to wikipedia, "Web 2.0", also known as the participative web, refers to all the techniques, functionalities and uses that have followed the original form of the web, www or *World Wide Web*. It is characterized by simplicity and interactivity (sociability).

This original form of the web is particularly concerned with interfaces and exchanges, enabling web users with little technical knowledge to appropriate new web functionalities. It gives Internet users the power to contribute, exchange information and interact. This makes the web surfer an active person on the web.

1.2.9.3 Internet user

The term "surfer" refers to any person who uses a web browser to visit web sites. From this visit, the surfer can obtain information on the net and also interact with other people.

1.2.10   Webmapping or dynamic mapping on the Internet

I. BALDE (2008) defines webmapping as a generic term that covers both the process of distributing maps via a network such as the Internet or Intranet, and displaying them in a browser. In other words, it can be called a web GIS. The discovery of web GIS through mapping services such as google map or google earth has been made possible by the rise of android.

1.2.10.1   Webmapping components

Webmapping requires three components: a client, a map server and geolocalized data.

G. AYEL (2013) explains the three components he used to implement webmapping.

- **Customers**

Cartographic clients are used to display spatial data. They can be either office software or cartographic clients running through an Internet browser. An API is used to provide a Javascript library and functions for integrating geographic data into a website. This programming language enables interaction with the user.

- **Servers**

Servers are storage devices that offer services to customers. They house the application and the data on which the user (client) can perform queries. There are Web servers (Apache, IIS), spatial data servers and map servers. A server makes it possible to carry out actions and other functions. Interaction with the server is facilitated by the PHP language.

- **Networks**

Networks are devices that enable communication between server and client via intranet, internet or geoservices.

1.2.10.2   Webmapping creation

According to G. AYEL (2013), to create an application for disseminating cartographic data, it is necessary to develop a spatial database. PostGis", "MySQL spatial" and "Oracle Spatial" are some of the most interesting ones. Map servers are also useful for distributing spatial layers. The most widely used are "Mapserver" and "Geoserver".

GeoServer is, as its name suggests, an OpenSource map server (like MapServer) that lets you distribute or modify spatial data on the web. There are other proprietary alternatives that provide more or less the same services, such as ArcGis Server, CubeWerx SDI Suite, ERDAS APOLLO ...etc.

Geoserver is developed in Java. V. Besand, L. Ecault (2020) state that Geoserver works on the server side as an application managed by a Java application server (servlet) like Tomcat. Geoserver has become the OGC reference implementation (model software) for data distribution according to the WFS and WCS (web service for vector and raster data, respectively) standards, and remains one of the best software packages for distributing maps (assembly of styled layers) in WMS.

1.2.11   Database concept

There are many theories on the concept of a database, depending on the area of interest. When it comes to representing the real world, the search for similarity often leads to the creation of models. On this subject, H. ABBADI (2012) believes that the real world is too vast and complex for our immediate and direct understanding. This is why we create "models" of reality that are intended to have a certain similarity with certain aspects of the real world. For knowledge of the nature of this reality, databases are created from these models.

According to M. BRASSEUR (2008), a database DB is an entity in which data can be stored in a structured way, with as little redundancy as possible. To share information, the database is often associated with a network. The resulting data is made available to different users, regardless of geographical location. Users are able to enter, consult and update data, all on the basis of access rights.

1.2.12   Database (RDBMS)

A. CORNUEJOLS (2010) defines DBMS as general-purpose software that handles the processes of defining, building, manipulating and sharing databases by and between different users and applications.

In an RDBMS, data is stored and linked. They correspond to attribute information describing the given space, while geographic objects are referenced geometric data. Attribute data are directly linked to a particular geometry in a GIS via ODBC computer connections.

I. BALDE (2008) explains that in an online design, maps are visualized by programs installed on map servers, which communicate via predefined protocols. Geometry is managed using the RDBMS spatial cartridge.

1.2.13   Data modeling

Spatial modeling consists in identifying and delimiting elements or objects in a geographical space, representing these elements by graphic units (points, lines, polygons, etc.) and distributing them over different layers. Once modeled, the data is stored in a GIS database.

The design of a database begins with conceptual modeling. It aims to identify the elements of the space under consideration, then determine the structures according to the characteristics observed. Finally, this process is completed by linking the various elements. This modeling phase takes into account both geographic and non-geographic data.

In this modeling process, entity structure is based on object attributes. Depending on its characteristics, an entity is structured into fields whose description is described in a data dictionary. According to J. C. OUEDRAOGO (2011), "the data dictionary defines the set of elementary data of interest to the field of study under consideration". In addition to the dictionary, there are also the different data models (conceptual, logical, physical).

- **Conceptual data model**

A conceptual data model is a set of concepts used to describe and manipulate real-world data, and the rules for using these concepts. Models are made up of 2 parts: a static part describing the data structure (MCD) and explicit constraints on the data (CI), and a dynamic part defining data processing (MCT).

The basic concepts of modeling are :

- Objects grouped into classes and identified,
- Links between objects with their cardinalities,
- Object properties,
- The multiple representation of objects. GAGER (2020)

- **Logical data model**

According to J-F. PILLOU (2008), the logical data model consists in describing the data structure used without referring to a programming language. It involves specifying the type of data used during processing. The logical data model therefore depends on the type of database used.

- **The relational model**

Each entity class in the conceptual model becomes a table in the logical model. The identifiers of the entity class are called *table keys*, while the standard attributes become table attributes, i.e. columns.

1.2.14   GIS portal

According to Esri France (2020), a cartographic portal is an intranet or internet site that offers a common gateway to a wide range of resources centered on the field of cartography.

In terms of information dissemination, Oracle (2005) defines a portal as a central location where any type of information can be made available to a wide audience. With the evolution of technology, many services using geographic data are now available to the general public via the web: map consultation, visualization, localization, search, etc.

F. MERRIEN (2011) states: Web 2.0 has enabled the initial development of geographic information on the Internet. But this development is set to grow further thanks to 2 other technological mutations: on the one hand, mobile terminals (smartphones, tablets...), and on the other, positioning systems, which use either satellite systems (Global Navigation Satellite Systems, including GPS) or the relay antennas of cell phone networks.

The author presents Web 2.0 in two dimensions:

- A social dimension: it is relational, oriented towards the general public, based on sharing and exchange between Internet users, who have become producers of information (Facebook, YouTube, Flickr, Twitter, Wikipedia, Google Maps...). In the field of geographic information, everyone can become a contributor, as in the OpenStreetMap project, which aims to create free maps of the whole world.

- A technical dimension: Web 2.0, based on AJAX (Asynchronous JavaScript And XML; a cocktail of pre-existing technologies), is applicative, i.e. its main technical contribution is to enable the creation of on-line computer applications, accessible with a simple browser, in which a fairly rich interface can be created.

  This interface is an HTML page, in which certain zones are used to exchange data with the server using XML (but also in other formats: JSON, HTML, plain text). AJAX can be seen as a federation of four elements: - a computer application on the server (coded in Java, PHP or other), 2/10 CGDD/DRI/MIG - an HTML page supplied by the server, displayed on the user's computer (client) and incorporating JavaScript code enabling the client to carry out certain simple processes without calling on the server, - data exchanges in XML (or other) between the server and the client, - CSS style sheets to display this data in a readable way on the client. Google Maps, Bing Maps and the Géoportail are computer applications that use AJAX technologies, in particular to display maps and make them interactive.

1.2.14.1 Portal functionalities

The functionality of a geoportal can be summed up as the provision of tools and data that enable users to extract the information they require. Various options remain available to users, including access and freedom of selection or choice of data, followed by manipulation, viewing, searching and querying. In addition to these functions, the portal can also have specific functions such as data addition, surveying and calculation tools.

To establish the functionalities of a web platform, it's necessary to assess the needs of its users. Who are these users?

On this question, J. C. OUEDRAOGO (2011) identifies two groups of users (those in the public domain and the administrator).

From a communication and visibility point of view, the first group is the main target to reach. And their needs are generally limited to data consultation and navigation.

The author goes on to say that, for public domain users, these are the standard functions for geographic data, i.e. :
- *navigation*: displaying, moving and zooming the map ;
- *Search*: information search based on geographic or attribute criteria;
- *export*: formatting of maps in printable format (e.g. PDF).

1.2.14.3   System architecture

On system architecture, R.P.TATSO (2011) explains: when a web client connects to the platform via the http protocol, it may request the display of a map corresponding to a specific geographical site. In this case, our platform retrieves the data linked to his request from the MySQL database, then sends the request back to the Google Maps server via the Internet, which immediately sends the corresponding map back to the client. The client can then interact with the map using client-side JavaScripts functions. To implement this architecture, we needed the following tools: MySQL data server (DBMS), Apache Web server, PHP scripting server - all contained in a single package.

1.2.14.4   Example of a GIS portal

The implementation of a geoportal depends on the objectives set out in the specifications. With regard to the functionalities described above, different platforms may have similarities. For example, some authors describe geoportals in these terms.

According to I. BALDE (2008), a portal can be organized as follows:

- A banner will be placed at the top of the screen. It will feature a logo and the names of all the free tools used to create the platform.

- The left-hand side of the screen is reserved for selecting the layers to be displayed.
- In the center, we have the display area for the map generated from the selected layers, as well as the palette of navigation tools.
- On the right we have the section dedicated to research.

In more detail, J. C. OUEDRAOGO (2011) describes a geoportal in four zones:
Zone 1 is the map display area. In its upper left-hand corner are the navigation tools for :
- To zoom in, left-click on the plus ;
- To zoom out, left-click on the minus;
- to move the map in one direction, left-click on the arrow indicating that direction.

Zone 2 displays the list of available layers. There are two (2) types of layers: base maps and overlay layers. The plus (+) and minus (-) signs in front of the labels allow you to unroll or close a node. To show or hide a layer, simply click on the option button in front of its name. This area has a button for closing it to free up space for enlarging the map.

Zone number 3 displays the legend for layers visible in zone number 1. If a layer is hidden, it does not appear in the legend. If a layer is made visible, it appears in the legend.

Zone number 4 displays the information search criteria. There are two (2) criteria: geographic and information type. The geographic criterion indicates that the search is based on the territorial unit selected.

1.2.15 Information

a. Definition of information

The Larousse defines information as any event, fact or judgment brought to the attention of a wider or narrower audience, in the form of images, text, speech or sound. In computing, it is defined as an element of knowledge that can be represented using conventions to be stored, processed or communicated.

For wikipedia, information means both the message to be communicated and the symbols used to write it. It uses a code of signs carrying meaning, such as alphanumeric characters.

In other words, information can be defined as an organized set of data about an object or event, facilitating its understanding. It is generally obtained by processing data. In practice, information plays a major role in the decision-making process, and in the understanding of phenomena on earth.

According to P.ROMAGNI & V.WILD (2012) information is considered to be "information that improves our knowledge of any subject".

b. Information distribution

Once the information is available, the question arises of how to disseminate it. Disseminating information is made easy and efficient by the Internet. It minimizes or even eliminates the impact of distance and geographical location on the transmission of information around the world. According to M. BEN HENDA (2012), the dissemination of information has been greatly extended by Internet tools and services. The HTML language is one of the new alternatives available on the network to contribute massively to this Information-Communication phenomenon. HTML is the lingua franca of the Web, and has become the unifying interface for virtually all Internet services.

1.2.16   Project visibility

The term "visibility" has become very popular in projects of late. It focuses on the large-scale knowledge of all the project's actions and achievements at a given point in its execution. To achieve this, several tools can be used.

1.2.17   Development Project Concept

A development project is a project generally financed by bilateral, multilateral or private donors, with the aim of improving a country's socio-economic level (GDP) and the living conditions and standard of living of the country's local population (GNP). J. G. KONYAOLE (2001)

According to the Project Management Institute (PMI), quoted by N. FONTIL (2009), a project is any activity carried out once, with a defined beginning and end, and aimed at creating a unique product or knowledge. It can involve a single person or thousands of people. It can last from a few days to several years. It may be undertaken by a single organization, or by a group of interested parties. It can be as simple as organizing a one-day event, or as complex as building a dam on a river.

In conclusion, this chapter 1 has presented the framework of the study and the internship (AgriFARM project). Next, the scientific context was extensively covered, providing an understanding of the various key concepts related to fields such as geomatics and its tools, RDBMS databases, the map server, interactive cartography and webmapping, information dissemination via the Internet, etc. This enabled a better understanding of the most appropriate topics on the subject of geoportal development for information dissemination.

The general introduction to the document presents the context and the state of the art of interactive mapping, especially at the local level. This problematic is followed by research questions, then hypotheses and finally objectives. Three specific objectives have been defined for this study, and a methodological approach is developed in the following chapter.

MATERIALS AND METHODS

The methodology developed in this chapter outlines the steps followed during field and laboratory work to create the geoportal. The data collection procedure and tools are detailed in the section dedicated to the georeferencing of AgriFARM project activities in Upper and Middle Guinea.

The laboratory section covers the entire process involved in creating the database (postgresql/postgis), processing field data, and acquiring and processing data from online (opensource) resources. The process of acquiring OSM, topographic and satellite image data is also covered. Data integration into the database and map server is also covered.

Following on from the laboratory work, this chapter concludes with the selection of the information to be disseminated, the drafting of specifications and contact with the service provider for the web development part of the geoportal.

In short, most of the content of this chapter contributes to achieving the specific objectives set out above. The following table provides an overview of the methodological approach.

1.3 Process overview

Research questions	Specific objectives	Assumptions	Section chapter 2	Indicators
Q1 - Can geomatics tools be used to map the activities planned for a development project?	ensure the geolocation/georeferencing of activities planned by the project throughout its intervention zone (Upper and Middle Guinea);	*H1*: geomatics tools can be used to geolocate, process, analyze and disseminate all project-related information.	Acquisition of cartographic data for the AgriFARM project	-a functional GIS based on 5 components made up of opensource solutions; the activities planned by the AgriFARM project are geo-referenced in the intervention zone (Upper and Middle Guinea)
Q2 - Does a geoportal platform ensure the dissemination of information on development project interventions?	Process and integrate data on AgriFARM project activities into a GIS database and map server for publication;	*H2*: a GIS database provides the information required for publication	GIS software Conceptual Data Model Database Map server	-A GIS database generated by an RDBMS (postgresql/postgis) is available and functional; -the data to be published are available on geoserver or as free files in shp or geojson format
Q3 - Does the availability of geographic information on the Internet contribute to the visibility of a project's actions?	make geographic data accessible on a web platform designed with classic dynamic mapping functionalities;	*H3*: Web programming tools can be used to design and distribute geographic	Creation of the web platform	-the data selected by the project coordinator are published on the geoportal; - http://www.geo-agrifarm.com

		information generated by GIS.		is accessible on the Internet via a browser;

1.4 Field methods

1.4.1 Project modeling

Analysis of the AgriFARM project's territorial approach has identified a spatial model based on several interdependent units. The project's area of intervention is essentially based on the geographical space occupied by the agroecological regions of Upper and Middle Guinea, made up of 15 Territorial Units of Economic Development (UTDE). These UTDEs are made up of 34 project target communes, within which markets, production basins, watersheds and tracks are located.

The project's activities are spread over the above-mentioned geographical areas. They include the development of sub-watersheds and production basins, market construction, the rehabilitation of rural tracks and sections of national roads, entrepreneurship, the establishment of cooperatives and other associations, and multi-stakeholder platforms.

The identification of areas or geographical units and the activities planned within them has led to a categorization of the project's basic data as follows.

- a. basic cartographic data (topographic)
    - administrative boundaries (country, natural region, prefecture, sub-prefecture, commune)
    - transport (national road network)
    - hydrography (ocean, river, watershed)
    - topography (contour line)
- b. sector data (directly linked to project activities and indicators).
    - breakdown of project zones (national intervention zone, Middle Guinea zone, Upper Guinea zone, UTDE, target communes, beneficiary localities, etc.)
    - production basin (plain, pond)
    - watershed to be developed (village land)
    - market
    - tracks and roads to be rehabilitated
    - villages affected (crossed by the tracks to be rehabilitated)
    - other project interventions (entrepreneurship, school fields, CVEP, etc.)

Data collection was carried out according to the project's operational planning, the level of progress of activities, and specific needs (reporting, map production and publication, etc.). Given the progress of project activities in the Covid 19 context, data collection activities focused on mapping the intervention sites planned in the project's conceptual document. Other parameters were used to characterize the intervention sites in terms of key indicators, phasing and financing.

1.4.2 Data acquisition

1.4.2.1 Field data collection

As the basic data (administrative boundaries, hydrography, topography, transport and land use) were acquired in the laboratory, field data collection focused on the project intervention sites. These are markets, rural tracks and national roads, production basins (plains and ponds), certain sub-watersheds and villages benefiting from the project's actions.

Data collection was carried out in accordance with a procedure in force (Monitoring & Evaluation manual), supplemented by a procedure relating to spatial data (georeferenced). Data collection tools were formalized to take into account the georeferencing needs of the data collection subject. A data collection sheet has been prepared for this purpose. It is designed to contain the location in longitude, latitude, altitude and characterization of the sites.

To carry out the collection activity, a series of field missions was organized in three periods, covering the entire project intervention zone. These took place in April in Middle Guinea, and in September in Upper Guinea, using project resources. A third mission was carried out in November 2020 to supplement the data from Middle Guinea.

The tools used are :

- GPS garmin map 64S

This receiver was used in remote areas where charging the android tablet was not possible to take waypoints and plot rural tracks. Depending on the area, it gave an accuracy of around 3 m. To operate, this gps receiver uses alkaline (AA) batteries. Depending on workload, a charge of two (2) batteries averaged 24 hours.

- Application gps essentials

It was chosen for its functionalities, notably the ability to secure data through export or online sharing, geo-referenced photo-taking, and accuracy often below 3 m (on Android techno). When connected to the Internet, it is also possible to view the data collected on the osm or google map platform.

To use it, we installed gps essentials on a techno android tablet via playstore. We then set up the positioning and measurement units.

- Qfield

It was used to verify and validate certain data in the field. The data is organized in layers in a QGIS project and directly accessible either by synchronization or manual transfer of the project to an android.

The application was installed via playstore (android). We then proceeded to select certain layout parameters on the android (display, scale, attributes, refresh map, etc.). To ensure correct positioning, the android location and qfield current position options were activated.

On QGIS, we prepared the target layers and a control point layer in a project, all saved in the same directory. This folder was copied to the android directory. The project is opened in qfield (settings, open project) and gives direct access to the layers (objects and attributes) for display, manipulation and modification.

As the entity addition tool was integrated, we used it to collect control points wherever necessary. Finally, the folder containing the project and layers was transferred to the computer for data processing and integration (QGIS). This application was also used to complete the linear alignment of rural tracks and sections of national roads.

Two (2) android samsung and techno tablets, chosen for their battery life, location accuracy (less than 3 m) and photo-taking capabilities. The use of Androids enabled access to online data such as OpenStreetMap and satellite imagery in the field. This was made possible by extensive GSM coverage of the study area.

The above-mentioned equipment, like the GPS and Android, is part of the project's acquisition for data collection. As for the QFIELD and QGIS applications, these are open-source solutions, so there is no cost involved in their acquisition. In terms of functionality, QGIS provides the same GIS functions as other commercial software (ArGIS, Mapinfo).

1.4.2.2 Calibration of collection tools

Before collecting data with any of the above-mentioned tools, we calibrated them to guarantee data quality (positioning). This calibration procedure takes into account aspects of measurement units and positioning.

For the gps essential application, from the parameter menu, we have made the following choices:

- distance units (kilometer, meter)

- position datum (World Geodesic System, 1984)
- position format (decimal for displaying coordinates in decimal degrees, UTM for meters).

For garmin map 64S gps, the initialization steps are :

- open the receiver's main menu, then the configuration option ;
- in configuration, select the parameters to be calibrated (System, Time, Units or Position format)

The following box gives details of each parameter:

System
- *GPS (Normal)*
- *Language (French)*
- *Battery type (Alkaline)*
- *Interface (Garmin series)*

Units
- *Distance (Metric)*
- *Altitude (Meters)*
- *Temperature (Celsius)*
- *Pressure (Millibars)*

Time
- *Time format (24 hours)*
- *Time zone (Automatic)*

Position format
- *Position format (e.g. UTM UPS for UTM)*
- *Geodetic system (WGS 84)*
- *Map ellipsoid (WGS 84)*

1.4.2.3 Data collection process

Once the GPS receivers had been initialized, we set up in an open area to ensure connection with the satellites. We waited for the receiver to pick up a minimum of four satellites before starting data collection.

So, for each point, we marked and gave a name and a useful description that aligned with the contents of the collection sheet. In relation to the lines, we programmed the route and covered the path (linear) of rural tracks and sections of national roads.

All along these tracks and roads, we took points of the villages crossed, crossing structures and other important data in establishing the baseline condition of these infrastructures.

For the market sites and production basins, we took location points to serve as reference points for the laboratory work. For this starting point, the contours of the plains and ponds making up the production basins were obtained in the laboratory following the classification of land use.

The linear layers (rural tracks and national roads) obtained in the laboratory from osm data were verified and validated in the field using the QField application. To achieve this, we prepared a project in QGIS and then synchronized it with QField. In the field, line trajectories were corrected as necessary.

1.4.2.4 Transfer georeferenced data to computer

- gps garmin map 64 S

For geographic data collection, the project uses the garmin map 64 S gps receiver. Part of the latest generation of garmin receivers, it has the advantage of having more functions than older versions such as the map 60. It has ports that enable it to function like a USB stick in terms of connection to a computer as a peripheral. It doesn't require any specific software or application for data mobility, making it easy for different users to access.

Data transfer to the computer takes place in three stages.

- Connect the gps to the computer using the USB cable (we recommend using the original cable, otherwise others are available on the market and can play the same role);
- From the explorer, find the device (Garmin GPSMap 64s). It contains the GPX folder (waypoints and tracks in current) whose files are organized in chronological order of point capture and saving. All waypoints taken in a single day are saved in a single file with the name waypoint followed by the date. This file, in gpx format, is transferred from the device to the computer, where it is then used directly in the qgis software.

To transfer data from the gps essentials application, we used two methods (internal storage and online sharing). From the export menu, we chose the format option (gpx 1.0, kmz, kml), the storage location or the sharing option (email, social networks, drive, etc.).

1.4.2.5 Surveys

For this project, the surveys had two objectives: firstly, to acquire data on the project and its intervention sites (key indicators, phasing, financing), and secondly, to identify the various issues and expectations linked to the implementation of a geoportal platform as an information dissemination tool.

A questionnaire addressing the various aspects of the subject was prepared and administered to a target audience internal to the AgriFARM project. Individual and semi-structured interviews were conducted with the project's technical team. These interviews were conducted on the sites in conjunction with the meetings and field missions carried out during this period.

1.5 Laboratory methods

With the exception of Envi and Google Earth Downloader, most of the laboratory hardware used in this project is open source. They offer the advantage of being freely licensed and easily accessible to users. Also, for this work, opensource resources have the essential functionality from data collection to dissemination.

1.5.1 Conceptual data model

Conceptual data modeling was carried out using the MERISE method. In view of certain difficulties encountered in using the free version of the JMERISE software, another free software, Analyse SI (v0.80 - cairns), was used for the present modeling.

Analyse SI has been downloaded from
https://sgbd.developpez.com/telecharger/detail/id/2578/Analyse-SI.
It offers the advantage of adding a large number of association entities, providing the data dictionary and generating, following the MCD, the MPD (Physical Data Model), the SQL version and the Logical Model of Relational Data (MLDR).

To create this MCD, a new document named "agrifarm" was created with the software. We then added each entity using the corresponding tool. A name is assigned to the entity and all associated fields, including the identifier. Note that for each entity, the ID will represent its primary key in the database.

After adding all the entities in the project, we identified and added the possible types of association between the entities. In general, associations are designated by verbs in the infinitive, e.g. compose, belong, ... Finally, we added cardinality links between entities.

a. Nomenclature of entities and fields

To meet GIS data management requirements, the nomenclature of the various entities or layers must be conventional. For this project, we took into account spatial coverage, project area and layer content as nomenclature criteria. These criteria enable us to situate the user in relation to the layer content.

For example, a layer is named as follows: xx_xxxx_xxxx
part1 : xx (layer coverage area)

part 2: xxxx (layer content)

part 3: xxxx (layer content or other relevant reference).

Example: gn_limit_prefc (limit of all prefectures in Guinea)

Fields have been named using abbreviations to keep the number of characters to a minimum. Associations were named with infinitive verbs, as required by the procedure. Given the similarity of certain associations, we used a verb several times, differentiating it by a number at the end, e.g. appartenir 1, appartenir 2. The cardinality link has been identified in all its forms (0,1-1,1-0, N-1, N).

The conceptual data model for this project is made up of 41 entities linked by 51 associations of different cardinalities.

b. Data dictionary

The dictionary is generated automatically after integration of the conceptual data model (CDM) using Analyse SI software. For each entity or layer, it gives the name of the fields and their individual characteristics, such as the meaning of the name, type and character size.

In addition to the data dictionary, Analyse SI automatically generated the physical model and sql code from the MCD.

From the MCD, we have extracted the sql code of a few entities. It takes into account three syntaxes, including mysql and postgresql.

```
DROP TABLE IF EXISTS gn_contour ;
CREATE TABLE gn_contour (Id_contour_guinee BIGINT AUTO_INCREMENT NOT
NULL,
nom_contour_guinee VARCHAR(30),
PRIMARY KEY (Id_contour_guinee)) ENGINE=InnoDB;

DROP TABLE IF EXISTS gn_rnat ;
CREATE TABLE gn_rnat (Id_rnat_region_naturelle BIGINT AUTO_INCREMENT NOT
NULL,
nom_gn_rnat VARCHAR(30),
pop_2014_gn_rnat INTEGER,
```

Id_contour_guinee **NOT FOUND**,
PRIMARY KEY (Id_rnat_region_naturelle)) ENGINE=InnoDB;

DROP TABLE IF EXISTS gn_radm_limit ;
CREATE TABLE gn_radm_limit (Id_radm_region_administrative BIGINT
AUTO_INCREMENT NOT NULL,
nom_region_administrative VARCHAR(30),
pop_2014_region_administrative INTEGER,
PRIMARY KEY (Id_radm_region_administrative)) ENGINE=InnoDB;

⍰

This sql code enabled us to create tables in a DBMS software, postgresql.

1.5.2 Database

1.5.2.1 Installing postgresql/postgis

For this project, we chose postgresql for its free license and its ability to handle spatial data through its postgis extension. The Posgrsql 12 version was downloaded from the official website www.posgresql.org/. During the installation process, we respected the service nature of postgresql under Windows, and created a password for the posgres superuser. The default port 5432 was maintained, as was the local encoding. The PostGIS version corresponding to postgresql 12 was installed using the Stack Builder application.

1.5.2.2 Database creation

The first step was to open the pgadmin 4 interface and connect to the server with the postgres superuser password. Then we created the database from the "Database" menu with the following options: name "agrifarm", owner "postgres", encoding "UTF8", connection limit "-1".

In order to manipulate spatial data, we added the postgis extension for the "agrifarm" database, using the "create extension" option by right-clicking on agrifarm. In front of the "name" section, we wrote postgis to access a drop-down list of which it is a part. Simply select it, then create. When updating the "agrifarm" database, we noticed that the "spatial_ref_sys" table was automatically created, taking into account all spatial reference systems.

After testing the database, we connected it to the GIS software (QGIS) to import the spatial tables. Note that most of the layers in this project were created and updated in QGIS, then exported to postgresql. However, it is possible to create the tables directly in the postgresql database, then add the entities in QGIS.

1.5.2.3 Creating and updating tables

The GIS database tables were created in two stages. All tables related to the country and project geographical information were created in QGIS GIS software. They were then imported into postgresql via postgis, as explained in the following sections of this document.

Other tables from the MCD have been created directly in postgresql. It should be pointed out that these remain empty (without data) because they are linked to project activities not started during the training period. This is the case for indicators, baseline status and activity progress. As a reminder, only planning data will be processed in this thesis.

1.5.2.4 Table creation in the postgresql database

With postgresql, we have two ways of creating tables (either from the sql editor or via the table menu in the public schema). In the table menu, we right-clicked to access the CREATE option, then Table. We then named the table for this example "gn_prj_ind_perimetres_irrigues" in the "public" schema, with "postgres" as its owner.

The second step consisted in creating the various fields or columns as set out in the project MCD. For each field, the data or character type was chosen, with length and precision parameters for decimals. Finally, the primary key was chosen for this case "id".

The second method is to use the sql command editor. You can prepare the text on notepad or enter it directly into postgresql, as illustrated in this box.

```
CREATE TABLE public.gn_prj_ind_perimetres_irrigues
(
    id integer NOT NULL,
    nom_perimetre character varying(30) COLLATE pg_catalog."default",
    sup_exploites integer,
    hommes integer,
    femmes integer,
    jeunes integer,
    menages_tres_vulnerables integer,
    CONSTRAINT gn_prj_bp_perimetres_irrigues_pkey PRIMARY KEY (id)
)
```

1.5.3 GIS software

For the choice of GIS software, we aligned ourselves with the AgriFARM project's policy of using opensource resources. We chose Quantum GIS from a list of opensource software such as gvSIG, uDIG and OpenJump. QGIS offers the following advantages, among others:

- compatible with grass and saga ;
- ability to use multiple spatial data sources and formats;
- easy connection to DBMS such as postgresql through postgis ;
- several extensions for specific tasks;
- connection to imaging platforms such as google and bing map ;
- available on android through qfield ;
- a large community of users;
- possibility of creating interactive maps.

The fact that the software is under constant development is a factor of stability. Bugs are often found in several versions.

1.5.3.1 QGIS installation

Version 3.6, compatible with the Windows operating system, was downloaded from the official website www.qgis.org. It was installed following the default step-by-step instructions.

QGIS is organized in the same way as traditional GIS software, with a menu bar, tools, explorer, drawing or map window and status bar.

Before starting to create the layers or tables, a qgis project named "projet agrifarm" was created to house all the layers that are generated. We followed the following link: Project menu - New. This enabled us to display and overlay all the layers.

At the same time, a number of project properties were defined, such as the Reference Coordinate System (RCS) in EPSG code 4326, corresponding to the WGS 84 longitude/latitude projection. The units of measurement for distance and area were also chosen, i.e. meters, kilometers and hectares respectively, in the "Project properties --- General" option.

1.5.3.2 Installing extensions

One of the special features of qgis is the ability to install and use extensions, which are developed according to specific themes. To install an extension, we followed the steps below:

- connect to the internet ;
- open the extension menu - install/manage extensions ;

- search for the extension name (e.g. openlayer plugin);
- select and install the extension.

1.5.3.3 Vector layer creation

We created vector layers (point, linear and surface) following the same approach, the only difference being the choice of object type and their characteristics (fields). The layers were created as follows:

- Layer" menu - Add a layer - New shapefile (shp) layer
- Layer name and location
- Select default UTF8 encoding
- Select geometry type (point/line/polygon)
- EPSG :4326 - WGS84
- Add a field: id is a field created by default by the system. For each entity type, as many fields as characteristics have been added. For each field, we've given a name in lower case, then the type according to nature and the number of alphanumeric characters.

1.5.3.4 Adding layers in qgis

- **GPS data**

Data collected in the field with the garmin 64 S gps receiver and the gps essential application were added to qgis as vector layers (Layer Menu - Add a layer -Add a vector layer). In the "Data source managers" interface, with a default choice of file as the source type, we searched for the target data in the vector dataset (browse the data directory and select the file type. The file type can be gpx, gps 1, kml, kmz.

Once the file has been opened, it gives the option of selecting the vector layers to be added, with a choice between waypoint, track and route. Depending on the layer content, we have selected waypoint (points), track (line and polygon). In this way, layers are displayed according to their content (objects).

Data collected by android via gps essential can be saved and then transferred in kml format, which is opened directly as a vector layer and then converted back to shp for any updates (structure and content).

To make any changes to these layers, we re-saved them in another layer, this time in shp format. With this format, all modifications were made (addition or deletion, field update, modification of graphic object shape).

- **Excel or text data**

As data collection was carried out on some sites by project managers, we received the data in Excel format. They were converted into csv format, separated by a semicolon. To project this table onto qgis, we followed the steps below:

- Layer menu - add layer - add a delimited text layer ;
- Load the file from its ;
- Give a new name or keep the old one;
- Select encoding (UTF 8)
- File format (check custom delimiters, tab and semicolon) ;
- Select field and record options (if applicable) ;
- Geometry definition (default point, set longitude and latitude fields according to x and y correspondence)
- Select geometry SCR (4326) ;
- Check spatial index

Once added, the graphic point objects are displayed in the drawing window. This saves the layer in shp format, making it easier to use. Note that Excel data can also be projected onto qgis using the "spreadsheet layers" extension. To access it, follow the qgis extension installation procedure.

- **OpenStreetMap data**

Collaborative mapping resources (OpenStreetMap) are the main source of basic topographic data for the project. These are supplemented by data collected at national level (Ministère du Plan, Agriculture de la République de Guinée, IGN, JICA).

These different data sources are complementary. Today's OSM data is updated over time and includes topography and administrative divisions. In addition, they present a country-wide land cover, facilitating access to vector layers on several themes (hydrography, transport, residence, etc.).

The acquisition of vector layers (point, linear and surface) via this platform has an advantage in terms of data overlay for a geoportal, as the imagery backgrounds of interactive maps use the same data (osm). This saves time in data collection and is a benchmark for any GIS project.

To access the OSM data, we followed the steps below:

- Connect to the Internet and open the https://www.openstreetmap.org/ link on a browser;
- Creating a user account and password ;
- Login and country search (Guinea) ;

- Depending on the objective, choose the standard map (admin division, land use, etc.) or the cycling map (topography);
- Export using the geofabrik download option https://download.geofabrik.de/
- Select continent, then country and file format (osm or shp).

1.5.3.5 Adding and processing OSM data in QGIS

Files in osm format are opened directly in QGIS for viewing and processing. The osm format involves extracting by entity type (points, lines, polygons), then saving each type as a shp layer

So we opened the osm file from the layer menu (add vector layer). Given the size of the area and the large number of graphic objects, the layer selection interface displayed all available layers. There is a very large number of entities, as land use at national level is taken into account.

By adding all these grouped layers, we have seen the display of vector layers: points (built-up areas, points of interest), lines (road network, watercourses) and polygons (administrative divisions, land use). To facilitate the extraction of entities by theme, the layers were saved in shp format.

The osm data contains enough information. Depending on the needs of the project, we made queries to extract the information. For example, for the polygon layer, the attribute table was visualized to understand the different fields. This enabled us to perform an expression-based query on the attributes to select all entities in the same category (e.g. administration). The entities selected as a result of the query are saved in a layer called "example administrative region".

The result of this exercise was the following layers: hydrography, road network, agglomeration, village and hamlet, built environment, administrative boundary, country boundary, land use. The osm data being universal, the fields were updated according to the data dictionary of the conceptual data model set up for this project.

The resulting layers were overlaid with field data for verification, correction and validation. This overlay was carried out on qgis as part of the vector layer update process.

1.5.3.6 Update vector layers s

It was applied to modify the structure of the layers and their content (graphic objects and attributes). It was at this stage that we processed and integrated all the data that had been taken in notes in the field on the one hand, and those taken from the project documentation (key indicators, planning, etc.) on the other. All these layers were adjusted and then superimposed on the base layers obtained from osm data processing.

- **Layer structure modification**

The osm data layers are made up of several fields, most of which we don't need for this project. In addition, the classification and division of certain entities, such as the road network, does not correspond to that officially used in Guinea. To make these corrections, the layers were renamed, then restructured according to the characteristics specific to the needs of the study.

Layer nomenclature is based on a conventional approach using abbreviations. For example, to name the boundary layer of Guinea's prefectures, we chose gn(guinée)_limit(limite)_pref(préfecture), hence the layer "gn_limit_pref". On qgis, to name a layer, right-click then save as to give the name and direct to the destination directory.

To delete or add a field in a layer, we performed this task on the attribute table. With qgis, as with any other sig software, any modification to a layer requires selecting the layer and switching to edit mode. Editing is present in the form of a tool on the layer's attribute table, along with adding and deleting fields and calculating fields. So, to add or delete a field on a given layer, we switched the layer to edit mode, which directly activated the tools linked to layer modification.

- **Add, modify and delete graphic objects**

We practiced this on several layers during the verification operation, correcting features such as roads, rural tracks, hydrography, administrative boundaries and land use. Advanced digitizing tools were used to add, move, modify lines, polygons (nodes) and delete entities.

For linear and surface entities, the snap option was activated to facilitate the joining of nodes for neighboring polygons sharing the same contour line (administrative boundary, land use). On the other hand, snapping enabled good line joins or intersections, particularly for networks (roads, hydrography).

On the osm data, after zooming in and superimposing on the bing map or google satellite imagery, we noticed breaks in certain features (lines). To correct this, we proceeded with digitization using the "add an entity" tool, which enabled us to complete the rural tracks and national roads after interpreting the image.

Before accessing this imagery platform on qgis via an Internet connection, the "openlayer plugin" extension was installed. This allows access to all online data (openstreetmap, google image, bing map, etc.).

In relation to the production basins represented by plains or ponds, the field GPS points were superimposed on the satellite image. Once the plain or pond had been located following interpretation of the land use, for which the GPS point was the main reference, the production basin boundaries were digitized.

The representation of the object must be faithful, which is why the anchor points or nodes have been corrected for digitizing failures. For a geoportal project, it is important to work on the representation of vector layer objects on the image. This ensures that features are correctly superimposed on the various map backgrounds (osm, hybrid, satellite).

The "node" tool was used to move and readjust the nodes of lines or polygons. This is an important step, as data quality depends on it. Finally, the geometry was checked and any necessary corrections made using qgis tools.

To separate or merge entities, the corresponding tools bearing the same names have been used in the processing of line and polygon graphic objects. They were particularly important in extracting sections of national roads and rural tracks.

Overall, this approach was very useful for processing the osm data as the basic layers of the study and complementing it with layers of production basins and other infrastructures. At the end of this process, the basic data (layers) of rural tracks, road sections to be rehabilitated and production basins were obtained.

- **Adding and modifying attribute data**

Attribute data are added to each layer's attribute table in edit mode. With qgis, this process is made possible by entering text or numbers in the corresponding record cells for each entity. Geometric fields (length, area, perimeter, longitude, latitude) are updated automatically.

- **Field update**

According to the different entities, linear and surface vector layers have been created, whose length, area and perimeter parameter values are linked to certain key indicators. This is the case for rural tracks, national roads and production basins. The field calculator tool was used on the basis of geometry functions dedicated to the various parameters, $x, $y (longitude and latitude), $length (length) and $area (area).

Upstream, in the properties of the project containing the layers, we chose the units of measurement for distance (meters, kilometers), area (hectares) and the display of coordinates in decimal degrees. In addition to geometric fields, the field calculator tool was also used to update other attribute data.

1.5.3.7 Validation of project base layers

At national level, the updated administrative division of communes was obtained from the Ministry of Planning. Data from the IGN and the Japanese cooperation agency JICA were used to verify land use. The layers of the administrative division of Guinea were validated following this triangulation of information from these different sources.

This validation was carried out in qgis, with all information sources superimposed and entities corrected using advanced digitizing tools. At a smaller scale of project intervention (watershed, production basin, village, etc.), satellite imagery was used to complement and verify the data. This saved time.

1.5.3.8 Satellite imagery

Google Earth and free satellite imagery (landsat, sentinel 2, SRTM) were used to support or complement field surveys. To achieve the short-term objective of locating the project's intervention sites, we used google earth and bing map images with qgis.

The use of google earth and bing map images is justified by the fact that the google earth application offers the opportunity to find more recent images than bing map. And the images are available over several years for analysis of the spatial evolution of the intervention sites.

After data from the main intervention sites (plains, rural tracks, etc.), we turned to topographical data. These provide contour lines and enable us to identify and validate watershed boundaries.

- **SRTM image**

For this study, topographical data are particularly useful for planning production basins and watersheds. They are also used to read the sites identified for the construction of markets, rural tracks and sections of national road.

Topographic data are available on NASA's earth data platform. STRM (Shuttle Radar Topography Mission) images with a resolution of 30 m are free of charge. They cover all countries in a grid pattern. They are available as raster images in tiff format.

To acquire these images, a registration step is necessary. For this project, we used Earthdata Login (nasa.gov). An account was created in the profile for the viewing and downloading process.

The srtm images were downloaded from https://dwtkns.com/srtm30m/. To do this, we searched for the study areas on this grid.

After downloading, the image named N11W012 (example) is added to qgis for processing. It is first visualized with other data from the area to ensure quality. We then proceeded to extract the contours with a line spacing of 10 m.

To extract the contour lines in qgis, we followed the steps below: Menu raster - Extraction - Contour (choose the interval between contour lines, e.g. 10 m). Note that following extraction, the m dimensions of the contour lines are generated.

In addition to contour lines, we delineated watersheds from srtm images. To do this, we worked on the image in qgis, but this time with grass and the r.watersed tool. We also verified the overlay of the hydrographic network available in vector layers.

For the project intervention sites, these images have provided topographical data, the delimitation of watersheds and watercourses, and the layout of certain tracks and roads. These data are essential to the project planning process, which is the subject of information dissemination.

Following on from the GIS project for AgriFARM, we have also collected free satellite images (landsat 8 and sentinel 2) for each territorial unit of economic development. These are particularly useful in the process of drawing up planning tools (watershed management plans, for example). It should be noted that the period covered by this brief does not include these activities, given the delay in starting up due to covid 19.

- **Landsat and sentinel 2 image**

Landsat and sentinel 2 images are an integral part of the GIS data component of the AgriFARM project. As such, we have proceeded to acquire them according to the following procedure.

- Registration via an online form on the https://ers.cr.usgs.gov/ website.
- validation of registration by email ;
- opening the site on a browser https://earthexplorer.usgs.gov/;
- search, zone delimitation (by polygon);
- choice of parameters (image capture period, coverage, cloud ...)
- select satellite from dataset option (e.g. landsat 8 or sentinel 2) ;
- display the result, then the footprint and the image preview (show browse overlay). We have chosen an image with an acquisition date of 22-09-2020.
- download options ;
- choice of last file (Level 1 Geotiff, data product, generally larger than other files).

After downloading, all the tapes are in a compressed file named LC08_L1TP_201053_20200922_20201006_01_T1. After decompression, the file contains all eleven (11) tapes.

Depending on the purpose, we will use strips B2 to B4 for land use (built-up area, cultivated area, vegetation) in the sub-watersheds. For ecological monitoring based on the assessment of vegetation stress, strips B4, B5 and B6 can be used.

- **Satellite image processing**

Landsat 8 images were processed using Envi 5.1 software. Initially, we defined the class typology for land use in a sub-watershed. As rural tracks, bush paths and watercourses were obtained from the osm data, we were interested in the built environment of villages, cultivated areas and vegetation.

On Envi, we compiled the bands to create a metadata. As the image scene generally covers an area larger than the study site, an extraction was carried out for this purpose. We used the layer staking tool in the basic tools menu to extract from the outline of the study area (as a vector file).

Before moving on to classification, bands B2, B3 and B4 were ordered according to red, green and blue (RGB) spectra. Finally, we performed an unsupervised classification using the Isodata option, then converted the class into a vector. Note that this classification must be completed by ground control (field GPS data). This results in a sub-catchment land-use layer (essential basic data for spatial planning).

1.5.3.9 Google Earth application

We downloaded google earth pro from https://www.google.com/earth/versions/. The setup was installed locally. Google earth pro was initially used to identify and locate sites such as plains and ponds. As part of this project, it was used to complete data on infrastructures (rural tracks and sections of national roads). It should be noted that the first data on tracks and roads were obtained from the Open Street Map platform.

To locate the zones on google earth, we carried out a search on the platform. Then, the GPS data collected in the field was transformed into a file format readable on google earth (kml or kmz). These are points and lines respectively locating plains, ponds, rural tracks, roads and watercourses.

Once localization was complete, we interpreted the area's direct environment using image interpretation keys (object shape, color, texture, contrast, etc.). To identify the contours of a plain, for example, we digitized the object's outline using the appropriate tools.

Rural tracks and roads were also either digitized (object completion) or modified to comply with the official layout. Modification was applied to rural tracks with several deviations. This is a triangulation of information between osm data, the national trails and roads directory and the reality on google earth.

The digitized or modified data were then processed in qgis and integrated into the GIS database, in particular for the perimeter (plains, ponds) and infrastructure layers (rural tracks and sections of national road). For use in qgis, the data were saved in shp format layers.

To complete the triangulation of information, we returned to the field with the data acquired from the Google Earth imagery. We carried out a ground check on certain areas where interpretation required confirmation. To achieve this, we used the qfield mobile application, which enabled us to find, confirm and correct the layers in the field.

As a continuation of our geomatics activities within the AgriFARM project, we will be using google earth pro to correct (verify) land use data for village terroirs as part of the development of sub-watersheds. A diachronic analysis of land occupation and use (e.g. dynamics of built-up areas and crops) would also be possible.

1.5.3.10   Data coding

To facilitate data queries (spatial and attribute), we applied a coding system aligned with that of the project's Computerized Monitoring System (CMS). This alignment served to harmonize data use within the project.

At national level, the codification was applied to the territorial division of Guinea.

- 2 characters (national) GN
- 4 characters (administrative region) GN01
- 6 characters (prefecture) GN0101
- 8 characters (sub-prefecture) GN010101

The current coding of the project's SSI corresponding to the zones is as follows:

- Project antenna (first letter of antenna name followed by number 10, e.g. L10 = Antenne de Labé)
- UTDE (corresponds to prefecture), (first letter of branch name followed by rank, e.g. L01 = Dalaba)
- Cu/cr (sub-prefecture/commune) (UTDE code followed by two digits, L0101 = commune of Ditin in the UTDE of Dalaba)

1.5.3.11 Integration of layers in the postgresql/postgis database

For this project, the spatial layers make up the bulk of the database. They were created in qgis as explained in the previous sections. They were processed and updated in qgis.

In this way, final version layers were exported from qgis to postgesql via postgis. To achieve this, the following procedure was followed.

- Layer" menu - add a layer - add postgis layers ;
- Click on "New" and fill in the name (agrifarm), host (local localhost), port (5432);
- Enter the user name (postgres) and password (created when postgresql was installed and used to connect to the server in pgadmin). You can save the password by checking store or save.
- Test the connection (a success alert appears temporarily at the top of the dialog box) ;
- Confirm with ok

Once the connection had been successfully established, we accessed the database from the "Database" menu, then DB manager. This gives access to the database manager, followed by the list of data providers (Postgis). With Postgis, all connected databases are displayed with their tables in the public schema.

As all project layers are saved in a local directory on the machine, individual layers were imported. In particular, the ID field was designated as the primary key for each layer. Other options were checked, such as the geom field (geometry), reference coordinate systems (SCR) and encoding (UTF8).

The imported layers were stored in the "public" schema of the "agrifarm" database.

Once the layers had been imported, pgadmin was launched with a connection to the server. We noticed that the "agrifarm" database had been updated.

According to this project's conceptual data model, all layers are linked together in such a way as to ensure database functionality (querying). To achieve this, fields are used as primary and secondary keys.

1.5.3.12   Map server (geoserver)

For data publication, we used the Geoserver map server. Geoserver has become the OGC reference implementation (model software) for data distribution according to the WFS and WCS standards (web service for vector and raster data, respectively), and remains one of the best software packages for distributing maps (assembly of styled layers) in WMS. (V. BESAND, L. ECAULT, 2020)

In addition, it features compatibility with postgis and the display of geographic layers in image form, with the option of defining styles in SLD format. With WFS, geographic data can also be displayed in vector form, providing the functionality for querying and modifying symbology and labeling. The modification of symbology and labeling is based on the thematic analysis performed on each QGIS product.

- **Acquisition and installation of geoserver**

At http://geoserver.org/download/, version 2.13.2 (stable) of the application was downloaded and installed locally (localhost) on port 8080. For access, an admin account and password were created. At the end of the installation, we ran a test on a browser with the url http://localhost:8080/geoserver/.

The sequence of steps on geoserver is as follows:

- Creation of a Workspace named agrifarm ;
- Creation of new warehouses (connection to postgis database) ;
- Adding new layers from the postgis database (agrifarm) ;
- Modifying layer styles (SLD file generated in qgis) ;
- Layer assembly for publication.

The use of the map server was interrupted following the first attempt to install and operate geoserver with the web platform development team. They encountered technical difficulties with their server. By mutual agreement, we decided to adopt a second method, given the financial, time and technical (development) constraints.

The next step was to transfer to them the geographic layers generated in the postgis database. It should be noted that the layers concerned are those selected by the project coordinators. After consultation with the project team, it was decided to align this first publication with the project's public information strategy. The content of the entire postgresql database will not be published. It will be published progressively to support the content of the project website's publications, on the understanding that this geoportal will be a full page of the project website.

1.5.3.13   Project map theme

Depending on the theme, geospatial analysis is an essential tool for using ISS data to produce maps. The choice of themes is based on project indicators and, above all, the information to be disseminated internally or to the public. The map production procedure is aligned with the project's zone code and indicators, organized thematically and technically materialized as a cartographic project.

The publication of AgriFARM project information through cartography follows four axes:
- programming of project activities: intervention area, phasing, financing, infrastructure, ...
- monitoring of project activities: baseline situation, periodic mapping of activities planned by UTDE, modeling of project dynamics;
- activity reports (quarterly and annual).

In 2020, priority has been given to programming activities. This has led to the mapping of all intervention zones, planned activities, infrastructures, financing and project phasing. On this basis, dynamic mapping through the GIS geoportal is aligned with the project's progress and the policy of publishing information to the general public. This leads to a selection process for information to be published via the web (general public).

The following table gives an indicative list of the map layers in the project's GIS database.

Table 1: Summary of mapped project data

Themes	Domain	Coverage	Layer name	Description

basic map data	administrative	national	gn_contour	Guinea border
	topography	national	gn_DTM	SRTM image of Guinea
	topography	national	gn_srtm_34_10	SRTM image of middle and upper Guinea
	administrative	national	gn_capital	Capital Conakry
	administrative	national	gn_radmin_limit	Administrative region boundaries
	transport	national	gn_rout_network	National road network
	administrative	national	gn_rnatu_limit	Natural region boundaries
	transport	regional	gn_rout_prin	National trunk road network
	transport	regional	gn_rout_second	National secondary road network
	administrative	national	gn_subref_place	Chief town of sub-prefecture (urban/rural communes)
	administrative	national	gn_subref_limit	Sub-prefecture boundaries
	administrative	national	gn_village_usaid	Village in Guinea (source usaid)
	topography	national	gn_elevation	Elevation in m
	protected area	national	gn_for_class	Classified forests, national parks, reserves, ...
	hydrography	national	gn_hydro	National hydrographic network
	topography	regional	gn_mg_curb_level	Contour (mid-Guinea zone)
	hydrography	national	gn_ocean	Limit of Guinea (Atlantic Ocean)
	administrative	national	gn_country_limit	Country bordering Guinea
	administrative	national	gn_pref_place	Prefectural capitals (cities)
	administrative	national	gn_pref_limit	Prefecture boundary
AgriFARM project data	watershed	regional	gn_prj_bv_zon	Watershed to be developed
	watershed	regional	gn_prj_bv_pts	

target municipality project	regional	gn_prj_com	Boundaries of municipalities (urban and rural) covered by the project
communes bordering the target project	regional	gn_prj_com_all	Municipal and intercommunal boundaries
		gn_prj_com_all_limit	
target municipality project	regional	gn_prj_cr_intercom	
project financing	national	gn_prj_financ	Distribution of backers by commune
Infrastructure to be rehabilitated	regional	gn_prj_infrst_march	Market sites to be built
Infrastructure to be rehabilitated	regional	gn_prj_infrst_tracks	Rural tracks and stretches of national road to be rehabilitated
LAND USE	national	gn_prj_luse_resid	Built-up area of the project zone
Production basin	regional	gn_prj_perim	Lowland spots, pond to be landscaped
	regional	gn_prj_perim_poly	Contour of plain, pond to be landscaped
phasing	regional	gn_prj_phasag	Annual project phasing by municipality
village affected by the project	regional	gn_prj_piste_villag	Villages crossed by rural track and national road infrastructures
location of project area	regional	gn_prj_places_resid	Villages and hamlets throughout the project area
Infrastructure of rural tracks and portions of national roads	regional	gn_prj_situ_pistes	Beginning and end of rural tracks and national roads to be rehabilitated
opened-up locality	regional	gn_prj_villag_pistes	Villages and hamlets crossed by tracks or

				portions of national roads
project headquarte rs	regional	gn_ucp_antenne	AgriFARM project offices throughout the country	
prefecture target project	regional	gn_prj_UTDE	Territorial Economic Development Unit boundaries (prefecture)	
target region project	regional	gn_prj_zon_inte r	Project area (Upper and Middle Guinea)	

Selecting data for publication

The implementation of dynamic cartography through a geoportal requires a selection phase for the information to be published. This is important for a project with an unlimited number of future users.

The publication of project data is governed by regulatory procedures in line with the policies of the donor and the Guinean government through the Ministry of Agriculture. According to the priorities of the moment, the selection is made by the monitoring and evaluation unit under the supervision of the project coordinator.

The selected layers were prepared in QGIS. This involved streamlining the layers, maintaining only those attributes that had priority for publication. They were then sent to the development team in a simple shp file format.

1. Creation of the web platform

Setting up a geomatics portal is a process made up of two major complementary aspects: geographic or spatial data and web development. The spatial data aspect of this project encompasses the entire process, from data acquisition to the provision of the GIS database to the developer.

This represents the GIS part of a geoportal which we were responsible for implementing. For the design and development of the web application, we called on the services of a specialist.

Based on advice from a former student of the Master GAGER class of 2019, we contacted a local startup specializing in web development. The company is Centre de Développement Informatique Guinée (CDI Guinée), based in T5, Commune de Ratoma, Conakry, Guinea.

For feasibility, we analyzed interviews with users (the project team), which revealed needs related to display, navigation, search, positioning and measuring distances and surfaces. As far as data is concerned, openstreetmap and satellite images were cited as cartographic backgrounds, in addition to national data.

Finally, project coordination focused on the content of the information to be published, which had to go through a selection and validation process. On this basis, a set of specifications was drawn up.

1.5.3.14   Design specifications for geoportail

For the design of this geoportal, specifications were drawn up and discussed with the development team. It served as a basis for discussion and reference throughout the development process.

The quotation was drawn up on this basis, leading to discussions on the financial value of the development service. Given the financial implications, the discussion of the content of the specifications led to an agreement to collaborate on the development of the geoportal.

As the geoportal is open to modification, we agreed on the main points of the specifications which are directly linked to the expectations of interactive cartography. This is reflected in the specifications document in Appendix 2. On a technical level, a mock-up was drawn up with a detailed description of its content. And geoportal navigation scenarios were described in algorithmic form.

- **Geoportal model**

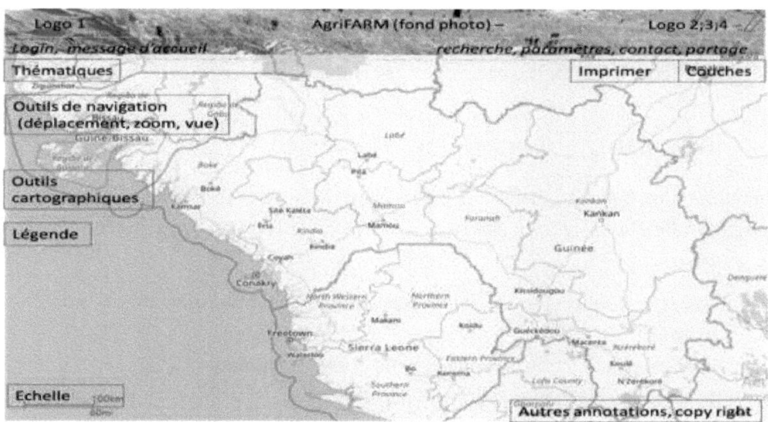

Figure 3: Geoportal model

Model description

1. Banner
-   Photo background
-   Logos (option 5, from left to right)
-   Various Annotations
-   Login, Contact, Home,
2. Map display area (OSM background with display area centered on Guinea outline)
3. To the left
-   Navigation tools
-   Geographic North (map feature)
-   Scale (map element)
-   Cartographic tools (spatial search, positioning and measurement of distance and area)
-   Legend (map element), moving window if possible
5. To the right
-   Layer control area (display or selection)
-   Print tool
6. Bottom of screen (horizontal orientation)
-   Other design and copyright annotations (e.g. SIG-AgriFARM 2020 in copy right)

Sliding panels are also provided to make room for cartography. As the background map is made of imagery, the transparency of layers, especially polygon objects, is set in the backoffice, as is the option to close all windows, whatever their size.

-   ***Geoportal navigation scenario***

The interaction between the actors (users) and the system is expressed by the navigation function. Internet access is a prerequisite for the following actions on the geoportal.

-   The user connects to the Internet and enters the geoportal address in a browser
-   The geoportal loads and displays its contents (home)
-   User clicks on the layer control panel
-   The layer list is displayed with a tab in front of each layer
-   The user selects by ticking the target layer tab
-   The system displays the graphic content of the corresponding layer as an image, point, line or polygon.

- User deselects a layer
- The system cancels the display of the layer content
- The user selects the zoom + tool, makes one or more clicks
- The system zooms in according to the number of clicks on the tool.
- The user selects the zoom tool -, then makes one or more clicks
- The system zooms out by the number of clicks on the tool.
- User selects move tool
- System moves tool from click point
- The user selects the move tool and clicks on an object in the layer
- The system displays the attributes of the selected object
- The user selects the measurement tool and then clicks on a point
- The system displays latitude/longitude coordinates
- The user clicks on a point with the measuring tool, then moves on to a second point
- The system displays the coordinates of the two points and the distance between them.
- The user selects the search tool, enters the name of a location
- The system displays and zooms in on this location.

1.5.3.15   Geoportal architecture

In the development phase, CDI Guinée proposed a multi-tier architecture for the geoportal infrastructure. This architecture is shown in the following figures.

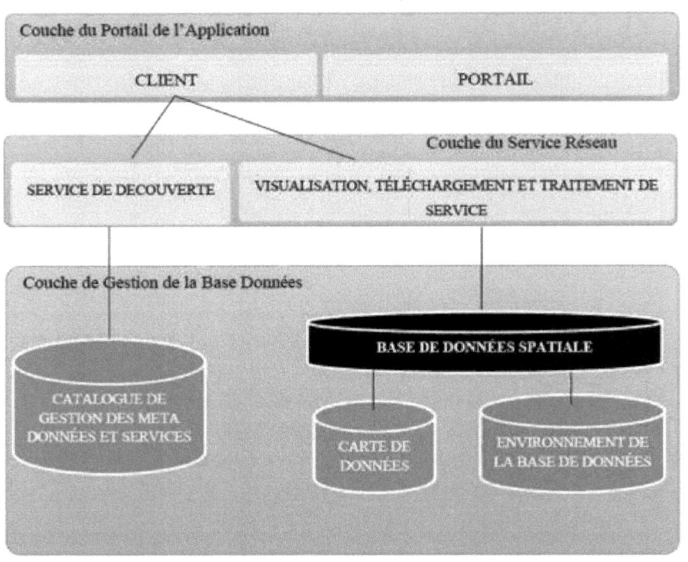

Couche du Portail de l'Application
CLIENT
PORTAIL
Couche du Service Réseau
SERVICE DE DECOUVERTE
VISUALISATION, TÉLÉCHARGEMENT ET TRAITEMENT DE SERVICE
Couche de Gestion de la Base Données
CATALOGUE DE GESTION DES META DONNÉES ET SERVICES
BASE DE DONNÉES SPATIALE
CARTE DE DONNÉES
ENVIRONNEMENT DE LA BASE DE DONNÉES

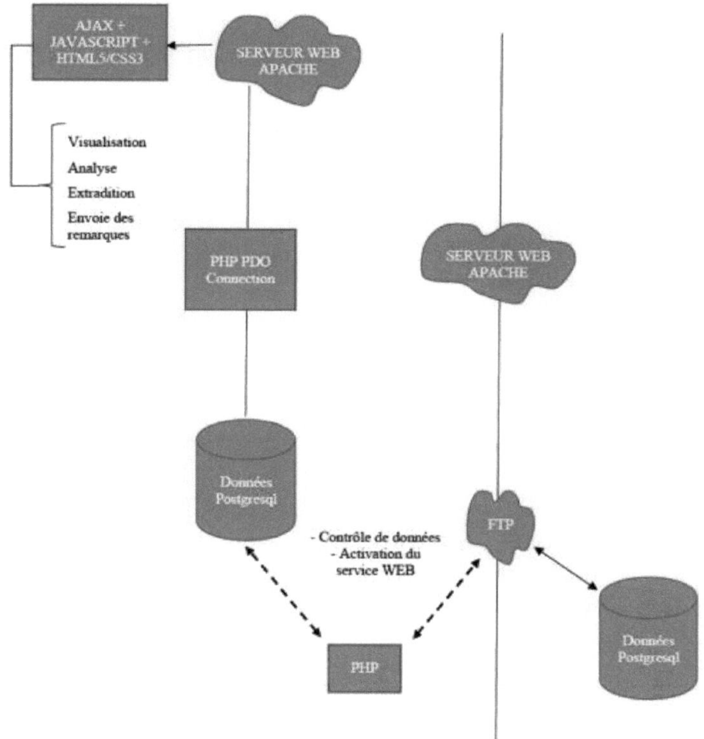

Figure 4: geoportal architecture

1.5.3.16   Interaction with the development team

Throughout the geoportal development process, we interacted with the team. We gave our opinion on the various design choices and ensured that the content of the specifications was respected. And when difficulties arose, we worked together to find solutions to keep the project moving forward.

On the technical side, we provided all the data to be published. The symbology was worked on in QGIS according to theme, and the color codes (hexadecimal) were shared with the development team.

After designing the first draft of the prototype, we carried out a functionality test. This continued during the process of adding tools and data. This process was concluded by sharing with the project team for advice and recommendations. Ultimately, the recommendations were taken into account before the geoportal was delivered.

In conclusion, this chapter has served as a framework for presenting the study, i.e. the AgriFARM project and the intervention zone. Knowledge of this area guides data collection through spatial modelling of the regions of Upper and Middle Guinea. Thus, the data to be collected are known, with their characteristics defined by the data dictionary established following the MCD.

From acquiring field data to making it available for web distribution, the whole procedure was explained, along with the various tools and hardware involved. These ranged from the simple Garmin GPS receiver to the Geoserver map server, GIS software (QGIS), MCD and postgresql database. The Google Earth, OSM and SRTM platforms were also used to acquire data from satellite imagery in particular. The project's cartographic repository was created using the OSM database.

Finally, in this chapter, the geoportal idea was translated into a mock-up, together with a set of specifications that served as a basis for discussion with the web development team contracted for the geoportal project. The result of the application of this methodology and its various tools is presented in the following chapter.

RESULTS, EVALUATIONS AND DISCUSSIONS

By applying the methodology developed in the previous chapter, we obtained the deliverables that contribute to achieving the overall objective defined for this study. The results related to the three specific objectives and their discussion are presented in this chapter. It presents the mapping of project activities, the GIS database and the geoportal itself.

The chapter concludes with a discussion of the results obtained, and recommendations and prospects for the study.

1.6 Mapping of project activities in Upper and Middle Guinea

For a GIS project applied to any field, geolocation is a key action in feeding the spatial database in support of indicators or characteristics specific to that field. As such, it serves as the basic tool for disseminating geographic information via static or dynamic (interactive) cartography.

1.6.1 Conceptual Data Model

Geolocation was carried out following project modeling, the result of which is the Conceptual Data Model (CDM). This CDM takes into account all the dimensions of the AgriFARM project throughout its life. It is designed to be dynamic. A total of 38 entities have been created and linked in accordance with the MERISE method and the project approach. To make the MCD easier to read, we have written a note explaining the abbreviations used in the nomenclature of entities and fields (table 2 and figure 4).

Table 2: extract from the data dictionary

Tables	Geometry _type	Field_name	Descriptio n	Type	Size
lessors_backgrou nd		Id	Donor identifier	INT_AUTO_INCRE MENT	
		name	name of lessor (acronym)	VARCH AR	
		rank	lessor's rank in financing	INTEGE R	
production_basin	point/polyg on	Id_bp	production basin identifier	INT_AUTO_INCRE MENT	

		name	name of production basin	VARCHAR	30
		category	production basin category	VARCHAR	30
		sup_plan	planned or target area	DOUBLE	5
		nbr_chp_school	number of fields farm school	INTEGER	
		benef	number of beneficiaries	INTEGER	
		benef_F	number of female beneficiaries	INTEGER	
		benef_J	number of young beneficiaries	INTEGER	
		etat_ref	reference status beneficiaries	VARCHAR	100
		exploit_act	area currently in operation	INTEGER	
		principal_speculation	main speculations	VARCHAR	100
sub_watershed	point/polygon	Id_bv	subcatchment identifier	INT_AUTO_INCREMENT	
		name	sub-catchment name	VARCHAR	50
		sup_plan	total surface area to be developed	INTEGER	
		sup_foret	area of forest to be managed	INTEGER	
		sup_tapade	area of tapade to be landscaped	INTEGER	
		sup_plantation	area of planting to be created	INTEGER	
		benef_agrof	number of agroforestry	INTEGER	

			beneficiarie s		
		benef_femmes	number of agroforestr y beneficiarie s (women)	INTEGE R	
		benef_jeunes	number of agroforestr y beneficiarie s (young people)	INTEGE R	
target_municipali ty	polygon	Id_com	commune identifier	INT_AUTO_INCRE MENT	
		name	official name of the municipalit y	VARCH AR	30
		pop	population according to the latest census	INTEGE R	
		nbr_distr	number of districts making up the municipalit y	INTEGE R	
		nbre_villag_prj	number of villages targeted by the project	INTEGE R	
		nbr_benef	target total beneficiarie s	INTEGE R	
		youth_target	target beneficiarie s young people	INTEGE R	
		target_woman	target beneficiarie s women	INTEGE R	
		nbr_perim	number of perimeters to be landscaped	INTEGE R	
		nbr_march	number of markets to be built	INTEGE R	
		pist_rout	linear of rural tracks to be	INTEGE R	

			rehabilitate d		
		company	number of businesses to be created	INTEGE R	
		nbr_chp_ecol	planned number of farmer field schools	INTEGE R	
		nbr_CVEP	planned number of Village Track Maintenanc e Committees	INTEGE R	
		sup_perim	area of production basin to be developed	INTEGE R	
		sup_bv	catchment area to be developed	INTEGE R	
		volume_credit_o ctroye	volume of credit to be granted (companies)	INTEGE R	

The following figure shows an extract from the conceptual data model (CDM) designed for the AgriFARM project's GIS database.

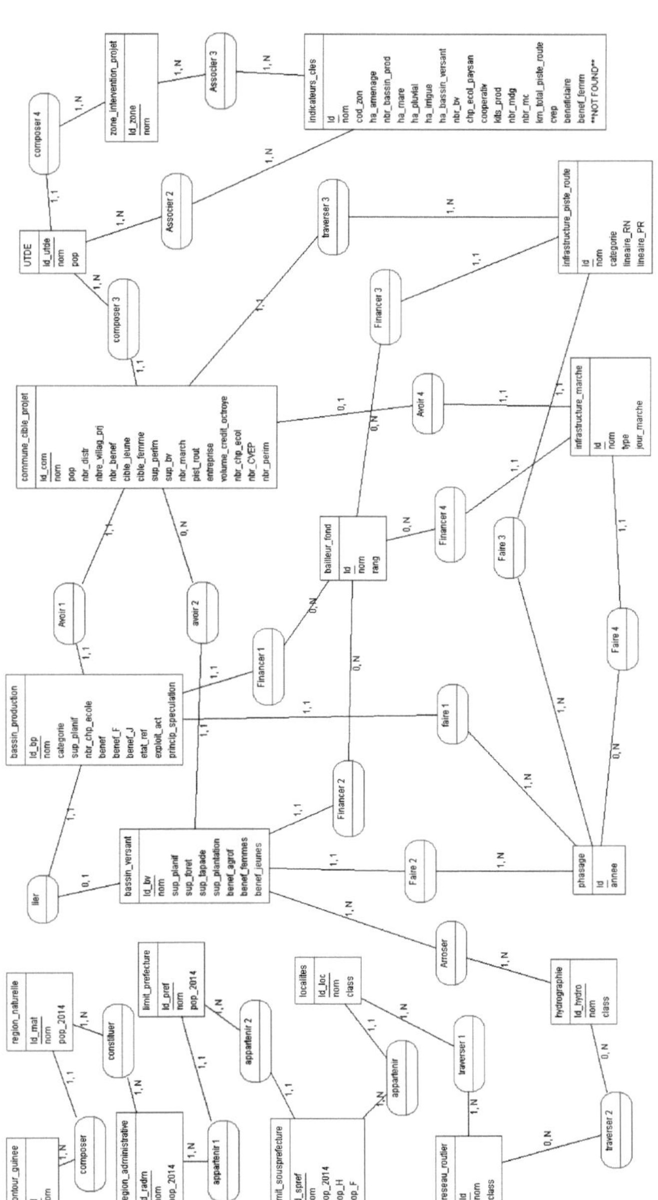

Figure 5: extract from the conceptual data model (CDM)

In this first year, given the progress (phasing) of the project in the field, the dissemination of information is focused on understanding and appropriating what is planned for the beneficiary communities. We have taken this aspect into account in the pursuit of the activities in this thesis, notably the geolocation of activities. Thus, planning data were collected, processed and integrated into a GIS database.

1.6.2 Georeferencing results

For this project, we developed a geolocation program for AgriFARM's interventions, based on the logic and territorial approach drawn from the design report. In addition, as AgriFARM's area of intervention is Upper and Middle Guinea, we collected basic cartographic data for Guinea.

These data have enabled us to better align the project's data with the national scale, making it easier to read the geographical situation of the project's interventions. In line with this objective, we collected the following data, organized in vector and raster layers.

A total of 39 vector layers were produced, with different graphic objects (point, linear and surface). It should be pointed out that some data, such as rasters and contour lines, were not created in the MCD, as they are used to produce other data. As an example, the illustration of some layers of the administrative division of Guinea are presented in this qgis project.

Figure 6: layers of Guinea's administrative divisions (prefectures and communes)

As far as spatial information at national level is concerned, we have mapped most of the basic data for Guinea. These were used to obtain data on the project's intervention zone. In the design phase of any project, the administrative division is used as the spatial reference frame.

A total of four (4) layers represent the administrative division of Guinea which, from macro to micro, are respectively the natural or agro-ecological regions, administrative regions, prefectures, sub-prefectures (commune). Cartographically, they are represented by surface graphical objects (representing boundaries) and are made up respectively of 4 entities (natural regions), 8 (administrative regions), 33 (prefectures), 341 (sub-prefectures) and one entity for the special zone of Conakry.

In addition, for prefectures and sub-prefectures, the chief towns are represented as graphic point objects, with the number of entities proportional to the number of polygons (boundaries). Mapping of the road and hydrographic network completes the basic data at national level.

1.6.3 Cartographic data from the AgriFARM project

1.6.3.1 Service area

Based on the project's territorial approach defined at the design stage, the project intervention area was mapped into two large zones (Upper and Middle Guinea) with reference to the administrative division and the targeted prefectures. Each zone (upper and middle) was subdivided into territorial economic development units (UTDE), the boundaries of which were mapped (15 entities). Finally, 35 target communes (urban, rural) were mapped. These are the gateway to the project and the subdivisions of the UTDEs.

To summarize the project's territorial approach in its intervention zone, three (3) layers have been created, representing the intervention zone (Upper and Middle Guinea), the UTDEs and the target communes respectively. For interventions straddling two or more communes, we have mapped all the intercommunalités. In addition, all other communes in the project intervention zone were mapped, making a layer of 151 entities. By way of illustration, the following figure shows the mapping of UDTEs and communes targeted by the project.

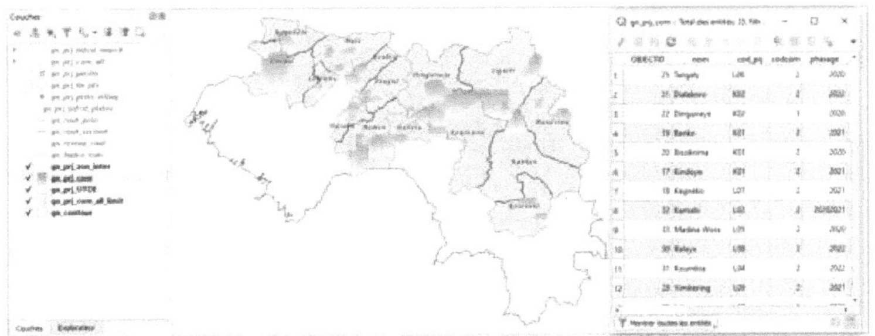

Figure 7: Vector layers for the project area (Upper and Middle Guinea, UTDE and communes)

Spatially, the Unité Territoriale de Développement Economique (UTDE) corresponds to the boundary of a prefecture, as shown in this figure (Mali prefecture). It is characterized by attribute data comprising the project's key indicators. A UTDE is represented in this figure.

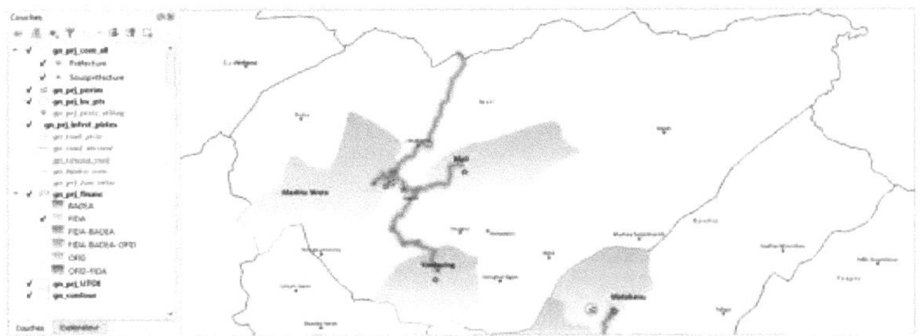

Figure 8: Project's territorial economic development unit (Mali)

According to the project's approach, a UTDE is made up of production basins and/or watersheds, and markets that are linked by road infrastructures to be rehabilitated. This is represented in cartographic terms in different layers on this figure.

In the field, mapping work has been carried out on production basins, sub-watersheds, market infrastructures, rural tracks and sections of national road. Project planning is now focusing on these elements.

1.6.3.2 Production and catchment areas

In the communes targeted by the project, there are production basins made up of plains or ponds. Around certain production basins, sub-watersheds have been identified and mapped, as shown in the figure below. Mapping involved the creation of point and surface vector layers. Depending on the objective, the point vector layer was used to represent these basins on a map at the national scale (Guinea).

The number of basins varies from one UTDE to another. We have mapped 23 production basins (22 plains and 1 pond) and 23 sub-watersheds. Two point vector layers represent the production basins and sub-watersheds. It should be noted that the entities in these layers were all characterized by attribute data supplied by the project's Monitoring & Evaluation System. They are of a planning nature (key project indicators). This figure shows the mapping of production and watershed basins in two layers.

Figure 9: Production basin and watershed vector layer

The sub-watersheds to be developed on this project as part of environmental efforts to protect lowland developments have also been mapped. Ultimately, this delineation along the watershed is an important planning step. It facilitates decision-making in the case of communes whose watershed sites had not been identified during project formulation. This is the case for the rural commune of Kankalabé. The figure below illustrates some of the sub-watersheds extracted from the SRTM image.

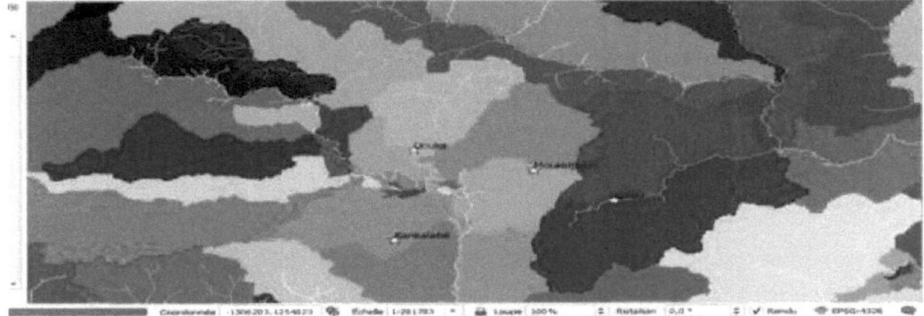

Figure 10: Extraction of sub-watersheds in the Tougué area (Kolou, Moukidjigué and Douka)

1.6.3.3 Infrastructure (markets, rural tracks or national roads)

Infrastructure activities include the construction of semi-wholesale markets (MDG) and collection markets (MC), as well as the rehabilitation of rural tracks and sections of national road. These rural tracks and sections of national road are being rehabilitated to facilitate access between production zones (production basins) and markets. Among the UTDEs, only Kankan and Mandiana are not involved in planning for the rehabilitation of rural tracks or national roads.

In the field, 23/24 rural tracks and national road sections were mapped. A linear vector layer was created for this purpose, with planning data as attributes. This includes the location of the infrastructure, i.e. the zero kilometer point (start or pk0) and the end point, name, category, phasing and funding.

The second part of the infrastructure concerns the construction of markets (semi-wholesale, collection). Twenty-six (26) market sites were mapped across the UTDEs, i.e. 21 collection markets and 5 semi-wholesale markets. They are represented as points in a vector layer as shown in this figure.

Figure 11: Mapping of infrastructure sites (market, rural tracks and sections of national road)

In addition, other geographical data were collected in the field along each rural track and stretch of national road. These included the villages crossed (directly opened up), details of the hydrographic network and contour lines.

In relation to villages and towns, we have mapped the center and built-up area, providing important information for locating the infrastructure. The following figure illustrates an infrastructure to be rehabilitated as part of the project planning.

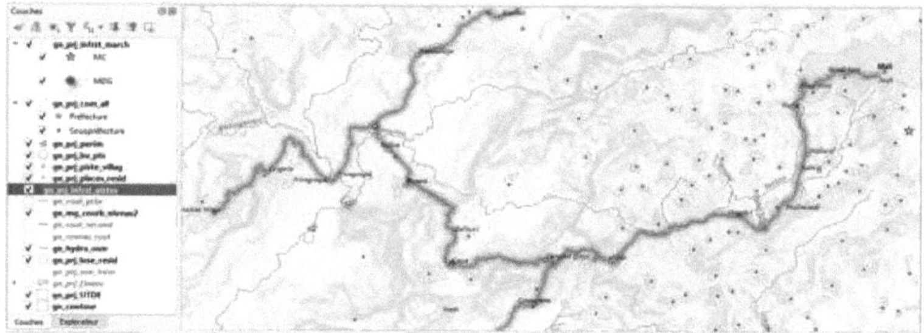

Figure 12: Mapping of national road section (Mali - Madina wora)

A total of 175 villages have been geo-referenced for the localities crossed by the road infrastructure to be rehabilitated by the project. The spatial location of these villages enables the infrastructure planning described in the DCP to be visualized. It is an important input for mapping rural tracks and sections of national road.

For each rural track or section of national road, the start (pk0) and end points constituting the infrastructure situation (itinerary) were mapped. The points of the villages crossed were organized in a vector layer and then characterized to facilitate the application of queries according to the infrastructure's situation. The following figure illustrates the position of the 175 villages and the road infrastructures that cross them.

Figure 13: Villages crossed by road infrastructure planned for rehabilitation in Upper and Middle Guinea

1.6.3.4 Financing and phasing

Being at the beginning of the project, we illustrated the financing and phasing of the project by mapping the 35 target communes (DCP). These themes were added to the surface objects in the vector layer as attribute data. For each entity (commune), data was added in addition to the name (the donor and the phasing or the year of the start of intervention in this commune).

For the purposes of symbology on this layer, funding has been taken as default in order to illustrate donors by commune. The project is financed by three donors, among others, who are represented here according to the activities planned for each commune.

Figure 14: Project financing by municipality

1.6.3.5 Project beneficiaries

The geographical distribution of project beneficiaries was carried out on the UTDEs, resulting in the creation of a vector layer with a surface object. Gender distribution (women, young people) was taken into account in the attribute data.

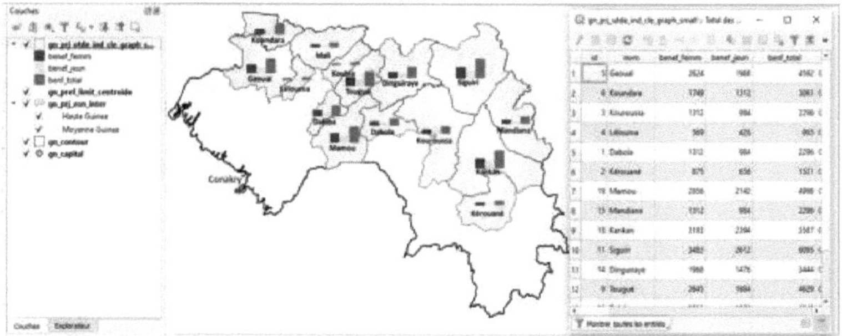

Figure 15: Mapping of project beneficiaries by UTDE

1.6.3.6 Topographic data

Among the topographic data, contour lines for the entire project area were acquired in linear objects of 10 m equidistance. Their attribute data are dimensioned points expressed in meters. They are used particularly in production basins and slopes. The following figure shows the topographic data for the Fatako, Tangaly and Koin study areas in the Tougué UTDE.

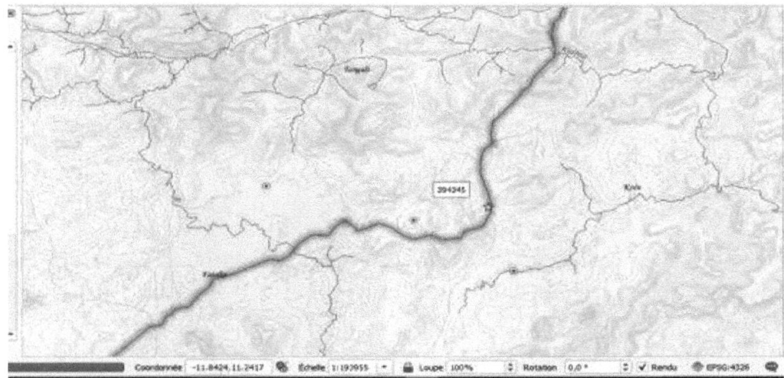

Figure 16: contour lines extracted from an srtm image of the Tougué area

The entire country's hydrographic network has been mapped, including the major watercourses (river, stream). Smaller watercourses such as marigots have been identified and digitized if they cross an infrastructure to be rehabilitated.

Details of the hydrographic network have been taken into account in the production and catchment basins. They are used to produce maps of these basins. We have therefore focused on the main watercourses required for project planning (dissemination of information throughout Guinea).

1.6.3.7 Other data

The project offices (1 Coordination and Management Unit, 2 regional offices) have been georeferenced in Mamou, Kankan and Labé respectively. A point vector layer is available with attribute data such as address and contact.

All these spatial data from the project are organized in the form of vector layers and integrated into a GIS database. This database is presented in the next chapter.

During fieldwork, we collected georeferenced photos. These were used to establish the reference condition of the various sites and infrastructures of rural tracks and sections of national road. A series of field photos is presented in Appendix 1.

1.7 GIS database

Once the data collected in the field had been processed, we proceeded to integrate them into the GIS database created for this purpose. As indicated in the methodology, the tables were processed in QGIS and then exported to the postgresql database via postgis.

1.7.1 Presentation of the GIS database

The GIS database is called "agrifarm" and is made up of 39 tables which are related to each other as shown in the MCD. It should be noted that all the tables related to project planning and presented above are integrated into this database. Spatial representativeness is ensured by graphic objects of the point, linear and surface type.

The other 9 tables are directly linked to the activity indicators not started by the project. They remain without entities for the moment. The figure below shows the database with a visualization of the project's Territorial Economic Development Units (UTDE) table.

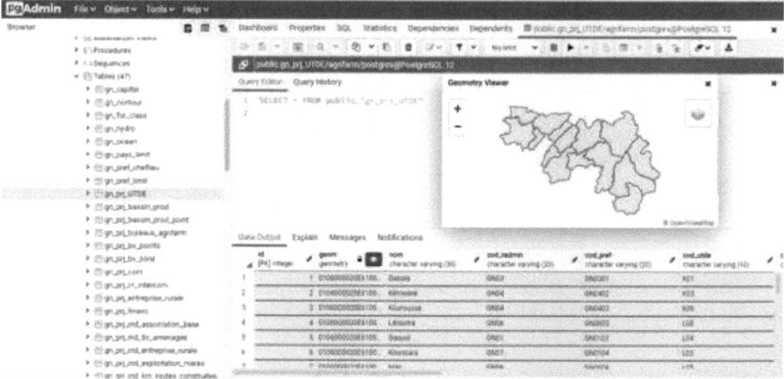

Figure 17: Display of the Territorial Units for Economic Development (UTDE) table in the postgresql database

Indicator data will be collected and integrated into the database as field activities are carried out. With regard to national data and project data in particular, we have made a selection according to communication priorities. These selected layers have been used in the production of static maps and also disseminated via the geoportal (interactive cartography).

3.2.2 Selecting data for publication

The data to be published in this geoportal fall into two categories, taking into account both spatial and thematic aspects. The first category is essentially made up of Guinea's basic geographic data (national coverage), and the second of data directly linked to the AgriFARM project.

Depending on the project's progress and communication strategy, the data to be published are selected by the monitoring and evaluation unit and the project coordinator. The following table shows the layers selected for publication, by theme.

Table 3: List of themes to be published on geoportail

Themes	Category	Coverage
Administrative	Country limit	National
Administrative	Agro-ecological division (natural region)	National

Administrative	Administrative division (administrative region, prefecture, sub-prefecture, urban/rural commune), district, village	National
Transport	Road network (roads, rural tracks)	National
Hydrography	Watercourses	National
Planning	Project area	Regional
Planning	Territorial Economic Development Units (UTDE)	Prefectoral
Planning	Project target communities	Sub-prefectural/communal
Planning	Project financing	Communal
Planning	Project beneficiaries	Prefectoral
Planning	Project headquarters and office	city
Planning	Production basins	Plain/mare
Planning	Sub-catchment areas	Communal or inter-communal
Planning	Infrastructure construction (feeder and semi-wholesale markets)	Communal
Planning	Rehabilitation of national roads and rural tracks	Intercommunality

As shown in this table, 5 layers represent the administrative division of Guinea (in boundaries and chief towns). From a boundary point of view, this division is made up of five levels (from the outline of the country to the rural communes). These layers have national coverage.

The project planning data is on a regional scale. They reflect AgriFARM's territorial approach and are represented by ten (10) layers. The project's key indicators are integrated as layer attribute data. In terms of attribute data, there was a further selection of information to be disseminated to the general public, hence the choice of fields for each table.

From a representational point of view, we prepared a project with all the above layers in QGIS. We applied a symbology and labeling, which allowed us to have a first visualization (especially for the appreciation and choice of colors to transmit in code to the development team).

The figure below illustrates these layers.

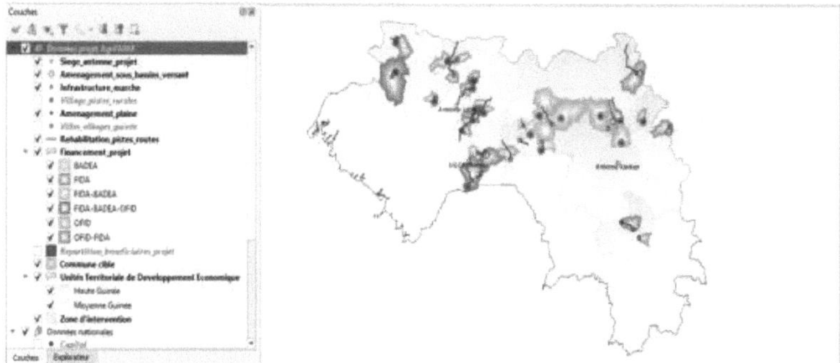

Figure 18: Draft layers selected for publication

This figure shows the AgriFARM project data layers selected for publication via the geoportal. They are mainly concerned with project planning.

Planning is spatially translated here by the location of the intervention zone at various scales in relation to the national territory (outline of the country). It is completed by information on the development sites (production basin and watershed), the infrastructure to be rehabilitated (rural tracks and sections of national road) and the infrastructure to be built (market).

Other information on the spatial distribution of project beneficiaries across the 15 UTDEs is also taken into account, as is the funding source. Donor distribution is based on commune level.

On a national scale, the information to be published is essentially administrative (country boundaries and administrative divisions). In the context of this project, satellite imagery and OSM will provide a wealth of other useful localization information. The figure below gives an overview of the national information to be published with the AgriFARM project.

Figure 19: layers of the administrative division of Guinea

The selected layers were then integrated into a map server for publication. As regards symbology and labeling, we used the style of each layer present in the above-mentioned qgis project. The following figure illustrates two layers published on geoserver.

Figure 20: view of prefecture and hydrography layers on geoserver

The difficulties encountered by the development team in relation to geoserver brought the publication of all data via the map server to a halt. Faced with this situation, and in agreement with the development team, we transferred the GIS database to be published. As a result, the vector layers in shp format were integrated into the geoportal.

1.8 Presentation of the geoportal

The geoportal set up as part of the AgriFARM project to improve the dissemination of planning information has an open-source geo-agrifarm.com domain. An extract of the source code is appended to this document (appendix 4).

Geo-agrifarm can be accessed via various browsers. From a design point of view, it has several parts, described as follows.

1.8.1 Home page

As its name suggests, this page presents the geoportal in all its components. As soon as you log on, OSM data is displayed, with a zoom on Guinea. It houses the banner.

The banner is the space dedicated to the publication of the project title, if applicable, and above all the logos. There are two types of logo: that of the AgriFARM project, currently on display, and those of our partners, including the Guinean government, IFAD, OFID and BADEA.

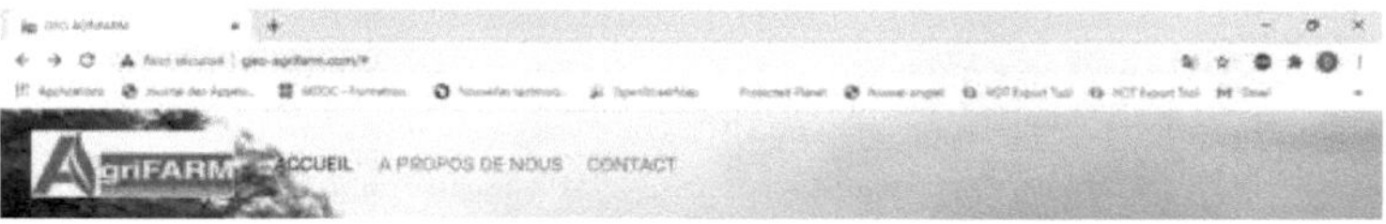

Figure 21: Geoportal banner

1.8.2 Private zone

It is associated with the back office, access to which remains the responsibility of the superuser. Functions such as account and password creation are carried out here. Other functionalities will be added in the future.

As the geoportal evolves, we'll be working to enable the creation of users who will be able to add geographic data (vector shp) that will be validated and authorized for publication by the administrator.

1.8.3 Public interface

This is the area dedicated to Internet users. Users can access the interface by entering the domain www.geo-agrifarm.com in any browser on a machine connected to the Internet. So, for this project, in view of the website's integration objective (access to the general public), there is only one group of users (visitors). They have access to the geoportal content (map, satellite image, plan) for display, navigation, searches, surveying and positioning, etc.). Other functions are also available, such as filtering and digital scaling, as shown in the figure.

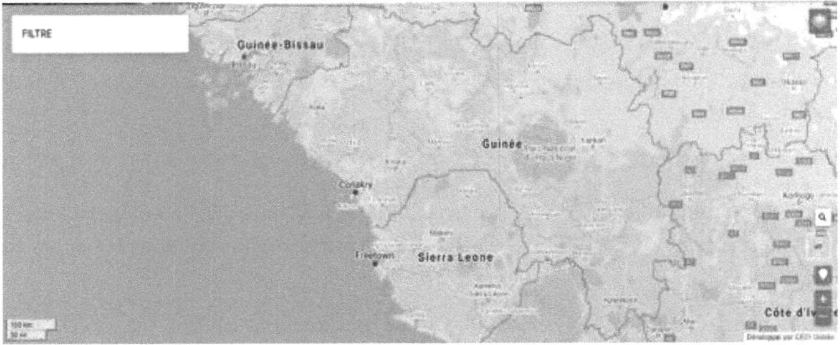

Figure 22: "geo-agrifarm" geoportal public interface

1.8.4 Tools

The functionality of the interface is based on the various tools deployed, in line with user expectations. These expectations are aligned with the classic operating principle of interactive cartography (display, navigation, search, positioning and surveying).

a. **Display**

Information is grouped into layers and organized in a menu. From the list, the user has the option of selecting the layer(s) he or she wishes to view. Note that only the map remains active by default, as shown in this figure.

Figure 23: View of geoportal information display options

b. Navigation

These include zooming in and out, moving and pointing. It should be noted that the user can use the external mouse for zooming and moving.

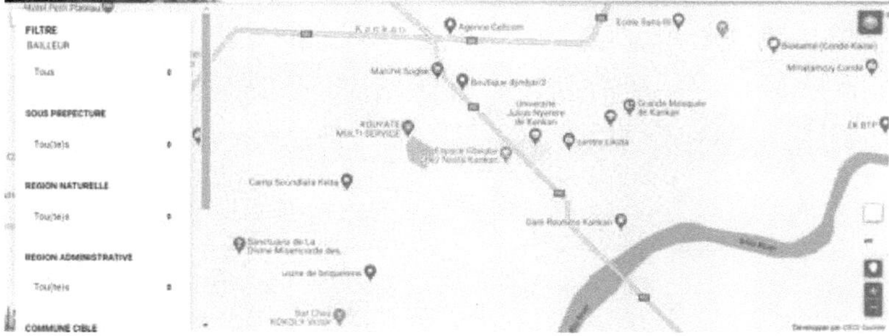

Figure 24: example of a zoom on the city of Kankan (osm)

The pointer can also be used to display attribute data for graphic objects via the tooltip.

c. Search

It is possible to localize the entire map (thanks to the google and osm backgrounds available on the platform). On certain project layers, we have also programmed a sorting system for rapid information retrieval. The figure below shows a sorting of IFAD-financed communes on the financing_project layer.

Figure 25: Search results by sorting IFAD-funded communes

d. Surveying tool

This is a single tool with a ruler icon. It displays geographic coordinates (longitude/latitude) by clicking on a position on the map. From two (2) points, it measures distance (in meters and kilometers) and area at three (3) points (in hectares and square meters). The figure below shows how to use the measurement tool.

Figure 26: Result of surface measurement at Moukidjigué pond, as in rural Fatako

1.8.4.1 Displaying attribute data

This operation is made possible by the tooltip. When the pointer is placed over a graphic object in a layer, the associated information is displayed in a small window.

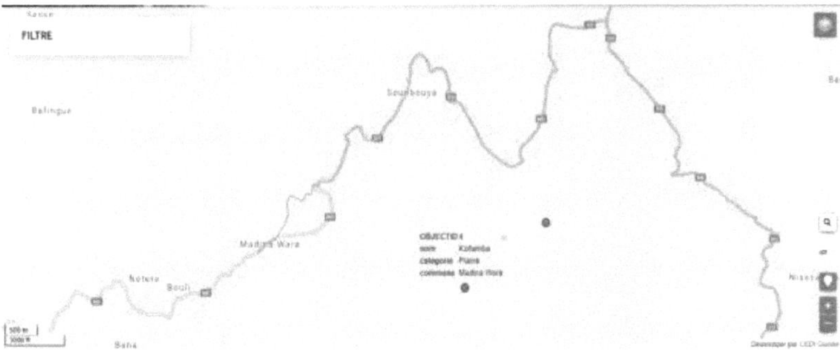

Figure 27: Tooltip on the Koitamba plain, Madina Wora commune

1.8.4.2 Published information (thematic layer display)

The information published to date on this platform relates to the map base and the themes. There are three (3) map backgrounds: map (osm), satellite image and hybrid. The following figures show each of the geoportal's map backgrounds.

Figure 28: map based on osm data (territorial view of Guinea)

Satellite imagery (from Google) forms the second cartographic background for this platform. It is available in two formats (simple and hybrid). The hybrid map has the advantage of integrating certain information from the google map (names of localities, main landmarks, etc.). The visualization of land use on satellite imagery is very practical, and enables the precise localization of information on the project sites. The following figure shows a view of Guinea on the hybrid image.

Figure 29: Example of a hybrid satellite image, Guinea

On the themes, the administrative division of Guinea is displayed on the figure below (prefecture boundaries). Thus, by activating the gn_pref_limit layer, Guinea's 33 prefectures are displayed on the map.

Figure 30: Display of administrative division at prefecture level

Then there's the project planning data, as previously announced. The following figure shows the infrastructure on a hybrid satellite image background.

Figure 31: infrastructure (market, rural track)

These two active layers display all the infrastructure (markets, rural tracks and sections of national road). The twenty-six (26) dots on this map represent the market sites to be built by the project. These markets are linked to the production zones (production basin and slope) by rural tracks and sections of national road to be rehabilitated. They are represented by lines on this map.

In the project's approach, production basins (plains and ponds) are directly linked to watersheds. In this figure, both types of information are displayed and superimposed in places, with the exception of isolated watersheds (with no production basin).

Figure 32: Information on production basin sites and watersheds to be developed

In terms of detailed information, this figure shows the infrastructure (sections of national road) to be rehabilitated by the project, the market to be built and the production basins to be developed in the Mali UTDE.

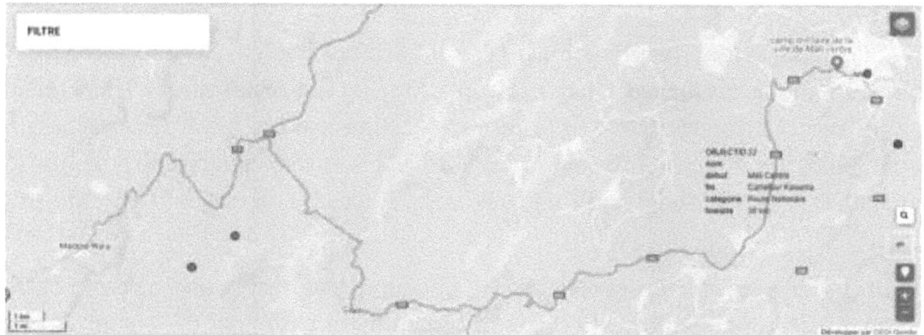

Figure 33: Mali market, Mali - Madina Wora national road and production basins

The geoportal offers the possibility of displaying and manipulating all satellite imagery information. This enables the web user to locate the direct environment of the planned infrastructure and, above all, to get an idea of the land use. The following figure shows the position of the site chosen for the construction of the collection market in the rural commune of Matakaou (UTDE of Koubia).

Figure 34: Sorting at Matakaou market, Koubia UTDE

Finally, the location tool gives the operator's position anywhere on the map.

1.8.4.3 Users

According to its objective, there is only one group of users for this geoportal. That is, Internet users. As the geoportal evolves over time, a second group could be created with the addition of new functionalities. This group will be able to add shp or geojson layers, which will be verified by the GIS administrator.

Assessment and discussion

Taking an overall look at the results presented above in their three (3) parts, we note with satisfaction that the objectives set have been achieved. The use of open source solutions has played a major role in achieving this result, and is a benchmark in GIS project management. As a result, decision-makers have direct access to geographic information linked to activities. This ensures that the potential of geomatics is better understood and taken into account throughout the life of a project.

On the other hand, the process of data acquisition, processing and quality control has enabled a good overlay of certain information, in particular the layout of rural tracks and sections of national road. This overlay was found to be in line with the data provided by the geoportal osm plan. This was due to the use of openstreetmap collaborative mapping data at the start of the data acquisition process for this project.

From a GIS database point of view, we have successfully connected the Postgresql database to the Quantum GIS software via the Postgis extension. This led to the creation of the GIS database presented above. The information integrated into the geoportal was supplied by this database.

The end product of this work is the geoportal, and its development was the fruit of the combined efforts of geomatics expertise in the very sense of the term (geography and IT). Its functionalities make interactive cartography available to web users.

In this way, it provides access to and manipulation of development project information. Access to geographic information for the general public is another aspect of this product, hence the process of selecting data for publication by the project coordinator.

To carry out this work, we followed a methodology established on the basis of literature dealing with the subject. Comparing the results obtained with the literature review, we observe a coherence in that the finished product, the geoportal, presents the functionalities of interactive cartography. A difference with the literature is noted at the level of map server use. This is due to the difficulties encountered by the development team.

Challenges and limits

Such work is not without its difficulties, but research has enabled us to find solutions. The case of the geoserver map server is a case in point. The IT specialists had encountered a problem with the server, and together we came up with a solution: we integrated the shp layers directly, using algorithms designed for this purpose.

So the fact that we are not using a map server is a limitation of this work. This is due to the fact that we worked remotely with the development team (covid 19 containment period). This caused difficulties in using geoserver in the geoportail platform hosting server.

Another important limitation is the fact that this work was carried out in two parts (geomatics and web development). This affected us in terms of operational planning for the completion of this work.

By acquiring these two skills, you'll have a better product when it comes to disseminating geographic information via the web, because cartography and web development are an art.

GENERAL CONCLUSION

In the context of this thesis, the work carried out has enabled us to achieve our main objective, "to set up a GIS portal". Relevant information from the AgriFARM project is disseminated through this portal, adding value to the communication strategy (increasing the number of communication channels). The dissemination of information is thus improved by precise localization and access to a larger number of users.

To achieve this platform, we first worked on setting up a GIS within the project. Throughout the system, priority was given to open source resources. They were used from the design of the conceptual model, data collection through to publication.

GIS needs data to function, so we acquired basic cartographic data for Guinea, then for the project intervention zone (Upper and Middle Guinea). From the project intervention zone, we proceeded with mapping based on the territorial approach (UTDE, target commune, intercommunality). Then, through field missions, we georeferenced all project activity sites by component. Production basins, watersheds, rural track infrastructure, sections of national road and markets were mapped.

All these data were integrated into a database created for this purpose. It was designed on the basis of the MERISE method and created in postgresql with its postgis spatial extension.

For the development of the geoportal, the team used the php/MapScript programming language for the dynamic display of web pages and maps, and apache as the web server. We used geoserver as our cartographic server.

As a product, we have achieved a geoportal for the www.geo-agrifarm.com domain that is functional on all browsers. Based on users' objectives and expectations, we have designed tools that align with the function of classic interactive cartography. These include display, navigation, search, positioning, distance and surface measurement functions.

These functionalities bring users closer to traditional office GIS services, with the possibility and independence of remote use via the Internet. It's also an opportunity to increase the number of users, regardless of geographical location.

This geoportal, set up for a development project in its start-up phase, will evolve over time to meet the need for improved information dissemination. Its source code remains open to modification. At this stage, it has enabled us to disseminate information on the project's planning for the 2020 start-up phase.

In view of the above achievements, we feel that we have achieved the main goal of setting up a geoportal to improve the dissemination of the project's geographical information. In addition to this role, this information dissemination tool enables users to use it as a workspace, by means of map backgrounds and satellite imagery.

With reference to the difficulties encountered during this study, we suggest :

- at the GAGER master's level, the inclusion of web development aspects in the content of training modules;
- at the GAGER master's program, to increase the time slots for the programming course and revise its content to adapt it to learners' future needs;
- to the AgriFARM project, the regular updating of this geoportal, adapting it to future needs with a view to transferring it to the Ministry of Agriculture and IFAD at the end of the project;
- development projects, the use of open source resources and capacity building for young people in the field of geomatics;

In terms of prospects, this geoportal has a dynamic character over time. In the short term, it will be integrated as a page on the AgriFARM project website. It will serve as an important support for the content of this website, especially in the project planning section.

In the medium and long term, it will be integrated into IFAD's GEONODE by 2022, and into the BSD of the Ministry of Agriculture, in order to perpetuate the project's achievements. IFAD's GEONODE is a digital platform for centralizing the geoportals of development projects in West Africa, for which IFAD is the main donor.

BIBLIOGRAPHY

1. Courses

Antoine DENIS, 2020. "Travaux Pratiques sur les Systèmes d'Information Géographique", Université de Liège (ULIEGE), Arlon Campus Environnement, Département des Sciences et Gestion de l'environnement, Unité Eau Environnement Développement (EED) http://hdl.handle.net/2268/190559

H. CHACKROUN, 2014 course Introduction to GIS and Remote Sensing in Civil Engineering, Ecole Nationale d'Ingénieur de Tunis, https://www.academia.edu/5138784/cours_SIG_syst%C3%A8me_dinformation_ g%C3%A9ographique_?auto=download

Michel TCHOTSOUA / Anaba Banimb R.C., 2019. GAGER MP 507 course. Geographic Information Systems: administration and integration of spatially referenced data.

Yacine KOUBA, 2018. Geographic Information System Course, Université LARBI BEN M'HIDI -OEB- Algérie Faculté des Sciences de la terre et d'Architecture Département de Géographie et d'Aménagement du territoire. https://www.researchgate.net/publication/324149696_Cours_de_systeme_d'infor mation_geographique

2. Master's theses

Hanae ABBADI, 2012. "Development of a geospatial information system using PostgreSQL and GeoServer for water data management in Morocco: the case of the Sebou watershed," Sidi Mohammed Ben Abdellah University, Faculty of Science and Technology.

David GADIOU, 2012. "GIS and complex data representation", Conservatoire national des arts et métiers, Information et communication pour l'ingénieur.

Issa BALDE, 2008. "Mise en place d'une plateforme de cartographie dynamique", Ecole Supérieure Polytechnique de Dakar - Design Engineer in Computer Engineering.

Jean Claude OUEDRAOGO, 2011 Setting up a webmapping interface on the "Capitalization of sustainable soil fertility management experiences in Burkina Faso",

Master 2 in computer science applied to geographic information systems, University of Douala

Guillaume AYEL, 2013. "Development of an interactive mapping solution to inform local residents about flood risk in Saint-Etienne," https://dumas.ccsd.cnrs.fr/dumas-00873081

Nolex FONTIL, 2009. "Projet de développement communautaire en Haïti : Méthodologie d'analyse des besoins locaux", Université Senghor d'Alexandrie, Master en Développement-Management de Projet. https://www.memoireonline.com/08/09/2623/m_Projet-de-developpement-communautaire-en-Hati--Methodologie-danalyse-des-besoins-locaux11.html

Philippe MARTIN, 2004. "Interface cartographique pour l'analyse territoriale multiscalaire de phénomènes sociaux", dissertation, Conservatoire national des arts et métiers, Centre régional rhônealpes center d'enseignement de Grenoble.

Olivier DUPRAS-TESSIER, 2016 presented the dissertation on "Conception d'un portail participatif d'information géographique web pour la mise en valeur des produits forestiers non ligneux d'après leurs potentiels de présence."

Renal Paul TATSO, 2011 "Intégration d'un observatoire urbain sur google maps : Cas des infrastructures de la santé de la ville de Douala", Master II thesis in Computer Science Applied to Geographic Information Systems, University of Douala.

3. Reviews and reports

Action contre la faim, 2017. activity report "Surveillance Pastorale Niger, November 2017" (www.Geosahel.info)

Bernier, E., Y. Bédard & M. Lambert, 2003. Cartographie sur demande sur le Web: de la personnalisation par couches cartographiques à la personnalisation par occurrences, Revue Internationale de Géomatique, Vol. 13, No. 3

Francis MERRIEN, 2011. Journée scientifique du Bureau des Longitudes " La nouvelle géographie " June 15, 2011 Intervention by, head of the geographic information mission at the French Ministry of Sustainable Development Internet and geographic information, http://www.geoinformations.developpement-durable.gouv.fr/fichier/pdf/IG_et_Internet-1_cle157b48.pdf?arg=177828603&cle=657a84eaced2229d5b97da5ccbe823740796846c&file=pdf%2FIG_et_Internet-1_cle157b48.pdf

Joseph Pascal MBAHA and Gille Baustert TCHOUNGA, 2020. "Webmapping and natural risk management: application to the Cameroonian coastline". Revue Espace Géographique et Société Marocaine (no. 33-34, April 2020).

Judith Gbêtowènonmon KONYAOLE, 2001. "Le suivi financier du projet de développement pendant son exécution ainsi que les procédures d'utilisation des

Fonds Banque Mondiale, PNUD, UNICEF" Ecole supérieure de commerce et de management d'Afrique.

ROMAGNI, (Patrick) and WILD, (Valérie) 1998. l'intelligence économique au service de l'entreprise, published by LES PRESSES DU MANAGEMENT, Paris, 1998, p.92.

4. Websites

Office québécois de la langue française, 2004, http://gdt.oqlf.gouv.qc.ca/ficheOqlf.aspx?Id_Fiche=8870689

Esri France, 2020, https://www.esrifrance.fr/SIG-vue-densemble.aspx

Sdigit, 2020, https://www.sdigit.fr/sig-cartographie/

Esri France, 2020, https://www.esrifrance.fr/portail-cartographique.aspx#:~:text=Un%20portail%20cartographique%20est%20un,le%20domaine%20de%20la%20cartographie.

Jeans Francois Pillou, https://www.commentcamarche.net/contents/661-merise-modele-logique-des-donnees, October 2008

Mokhtar Ben Henda , http://www.benhenda.com/hdr/documents/Manuels/2000_DUTICE_UTICEF/uv1_a/diff_outils.html, 2019

David Galiana, https://www.planzone.fr/blog/ameliorer-visibilite-projet, January 2018

WorlPress, https://lesdefinitions.fr/information, April 2011

Larousse, https://www.larousse.fr/dictionnaires/francais/information/42993, 2020

Guy Bonnerot et al. Encyclopedie universalis France, https://www.universalis.fr/encyclopedie/cartographie/, 2020

Le Quebec Géographique, https://quebecgeographique.gouv.qc.ca/education/geographique.asp, 2020

Laval University, https://www.scg.ulaval.ca/la-geomatique-cest-quoi, 2016
Journal DU Net, https://www.journaldunet.fr/web-tech/dictionnaire-du-webmastering/1203369-geolocalisation-definition-applications/, May 2020

Quantum GIS, http://www.qgis.org/fr/site/forusers/download.html, 2019

World Bank, https://www.banquemondiale.org/fr/country/guinea/overview, 2020

APPENDICES

Appendix 1: some images from the field

Photo 1: Bowoulena plain, rural commune of Banko, (latitude 10.619248, longitude -10.691558)

Photo 2: Safé stream (Safé village, Dinguiraye - Kaboukariah rural track), Dinguiraye urban district, (latitude 11.20491, longitude -10.671436)

Photo 3: Kaboukariah exit, Kaboukariah - Dinguiraye rural track, Cisséla rural commune, (latitude 11.01505, longitude -10.52568)

Photo 4: Berteya - Malikiya rural track, Mamouwol stream, Soyah rural commune, (latitude 10.12119, longitude -11.92576)

Photo 5: Gaya national road - Kaouma crossroads, Borko, rural commune of Gayah, (latitude 12.03448, longitude -12.43260)

Photo 6: Kamabi - Saréboido rural track, Kangueka stream, Kamabi rural commune, (latitude 12.39043, longitude -13.44767)

Appendix 2: specifications for the design of the geoportal - AgriFARM

Order: Design of a portal displaying interactive cartography of the AgriFARM intervention zone and associated documentary or multimedia content.

Order overview

- administration interface (back office) for adding content at a later date.
- Objects with point, line and surface geometry
- generic background map (OSM, Google satellite/hybrid/map)
- User management (granting access rights)
- Interactivity (movement, zoom, pointer, display, view)
- PDF map printing
- Use of cartographic tools

Cartography gives you the opportunity to :
- display of attribute data and pop-ups linked to clicked objects (points, lines or polygons).
- It's up to the user to interact with the display of layers (some layers are permanently displayed, so the user can display the desired layer without affecting the others).
- Locate (integrate location tool)
- Possibility of displaying geographic coordinates (cursor position on map)
- Navigation
- Use of cartographic and search tools

1. ***The back office of the cartography and admin module includes the following options:***
- Add, replace, update or delete map layers
- Layer display definition (mandatory display layers that cannot be deactivated by the user)
- Enable custom symbology (SVG)
- Enable default display of OSM background map
- Select the fields to be displayed for each layer
- User management
- Tool management
- Set up different zoom levels and good image resolution
- Transparency or opacity adjustment option for a transparent layer
- Map background (OpenStreetMap active by default)

2. ***Right of way***

Outline of the Republic of Guinea

3. ***Information on map layers***

- Objects (point, line, polygon) type of layers to be added example city location point, roads, prefecture contours
- Objects associated with attribute data
- Shp layer format or any other format from geoserver, for example
- UTF-8 encoding
- EPSG 4326 projection

4. *Themes*

In the back office
- Add, update and delete themes.
- Above all, allow layers to be renamed to make them easier to understand for Internet users.

Example

Administration

    - Regional boundary (natural and admin)
    - Prefecture
    - Sub-prefecture

Project planning (layers are grouped by category or theme. In this case, planning is a theme that includes the layers phasing, financing, utde,....)

    - Phasing
    - Financing
    - UTDE
    - Target municipality
    - Intercommunality
    - Location
    - Roads and tracks to be rehabilitated

Activities

Indicators

Impacts

Effects

5. *Cartographic backgrounds*
- OpenStreetMap, satellite and hybrid images (bicycle type)

6. *Navigation and display tool*
- Moving
- Zoom (-+)
- pointer
- opacity or transparency setting
- view (front, rear)

7. *Mapping tools*
- Display XY coordinates (in decimal degrees)

- Measurements (distance, area)

8. *Map elements*
- North
- Graphic or digital scale
- Legend (selected layers)
- Logo and copyright AgriFARM

9. *Requests*
- Query capability (location search)

10. *Printing* (optional)
- in pdf
- integrate the option (Card title) the first time you click on the print icon

11. *Share*
- Share with facebook, twitter, email (optional)

12. *Users and access rights*
- Public

13. *Properties and rights*

14. *Contact*

Appendix 3: Questionnaire overview

Q1 - What is the logical framework for this project?

Q2 - What are the project's areas of intervention? Present them in order of importance and specify the relationships between the areas.

Q3 - What are the project's various activities and how are they distributed geographically?

Q4 - What are the indicators linked to project activities?

Q5 - What is the geographical breakdown of project beneficiaries by type of activity?

Q6 - Is there a project monitoring and evaluation manual? If so, how is geographic information integrated?

Q7 - Is the monitoring and evaluation system computerized?

Q8 - What planning and monitoring tools are used?

Q9 - What decision-making tools do you use?

Q10 - How is information disseminated?

Q11 - How do you present the project's activities and results?

Q12 - What tools can be used to present the project's visibility?

Q13 - What GIS skills does the project or department have?

Q14 - Is there a web platform for the project? If so, how do you present the project's dynamics?

Q15 - How can GIS help disseminate information quickly and efficiently?

Q16 - What is the history of GIS use?

Q17 - What is your perception of the contribution of GIS to the dissemination of information?

Q18 - What data do you think should be included in a project GIS?

Q19 - What are your expectations in terms of communication and dissemination of information on project activities?

Q20 - What functionalities and themes would you like to see on the project's webGIS platform?

Q21 - How important do you think a GIS platform can be in the process of disseminating information to make project actions more visible?

Q22 - Have you used interactive mapping via the Internet? If so, what are your impressions?

Q23 - What are your expectations of the project's GIS (themes, functionalities, dissemination of information, etc.)?

Q24 - Why is it relevant to set up a GIS platform for the project?

Q25 - How do you want to access project information?

Q26 - Do you have an idea for an innovation to improve the dissemination of information about a development project?

Appendix 4: geoportail source code extract

Javascript :

```javascript
//Satellite
var tiles =
L.tileLayer('https://server.arcgisonline.com/ArcGIS/rest/services/World_Imagery/MapServ
er/tile/{z}/{y}/{x}', {
  maxZoom: 20,
  minZoom: 7
});
// Streets
var googleStreets = L.tileLayer('https://{s}.google.com/vt/lyrs=m&x={x}&y={y}&z={z}',
{
  maxZoom: 20,
  minZoom: 7,
  subdomains: ['mt0', 'mt1', 'mt2', 'mt3']
});
// Hybrid:
var googleHybrid =
L.tileLayer('https://{s}.google.com/vt/lyrs=s,h&x={x}&y={y}&z={z}', {
  maxZoom: 20,
  subdomains: ['mt0', 'mt1', 'mt2', 'mt3'],
  minZoom: 7
});
// Satellite:
var googleSatellite = L.tileLayer('https://{s}.google.com/vt/lyrs=s&x={x}&y={y}&z={z}',
{
  maxZoom: 20,
  subdomains: ['mt0', 'mt1', 'mt2', 'mt3'],
  minZoom: 7
});
// Terrain
```

```javascript
maxZoom: 20,
  subdomains: ['mt0', 'mt1', 'mt2', 'mt3'],
  minZoom: 7
});
var map = L.map('map', {
  center: [9.934886500000001, -11.283844999999985],
  zoomControl: false,
  zoomAnimation: true,
  zoom: 7
});
// .fitBounds([
//   [7.132612217968886, -17.402992633084757],
//   [14.012596641868884, -4.044774700548996]
// ]);
var highlightLayer;
function highlightFeature(e) {
  highlightLayer = e.target;
  highlightLayer.openPopup();
}
map.addLayer(googleStreets);
map.attributionControl.setPrefix(
  'Développer par <a href="https://cediguinee.com">CEDI Guinée</a>'
);
var autolinker = new Autolinker({
  truncate: {
   length: 30,
   location: 'smart'
  }
});
```

```javascript
L.control.locate({
  position: 'bottomright',
  locateOptions: {
    maxZoom: 19
  }
}).addTo(map);
L.control.scale({
  updateWhenIdle: true,
  imperial: true,
  metric: true,
}).addTo(map);
var measureControl = new L.Control.Measure({
  position: 'bottomright',
  primaryLengthUnit: 'meters',
  secondaryLengthUnit: 'kilometers',
  primaryAreaUnit: 'sqmeters',
  secondaryAreaUnit: 'hectares'
});
measureControl.addTo(map);
document.getElementsByClassName('leaflet-control-measure-toggle')[0].innerHTML = '';
document.getElementsByClassName('leaflet-control-measure-toggle')[0].className += ' fas fa-ruler';
var osmGeocoder = new L.Control.Geocoder({
  collapsed: true,
  position: 'bottomright',
  text: 'Rechercher...',
  title: 'Saisir votre mot-clé'
}).addTo(map);
document.getElementsByClassName('leaflet-control-geocoder-icon')[0].className += ' fa fa-search';
```

```javascript
var bounds_group = new L.featureGroup([]);
function setBounds() {
  map.setMaxBounds(map.getBounds());
}
function style_contour_guinee_1_0() {
  return {
    pane: 'pane_contour_guinee_1',
    opacity: 1,
    color: 'rgba(35,35,35,1.0)',
    dashArray: '',
    lineCap: 'butt',
    lineJoin: 'miter',
    weight: 1.0,
    fill: true,
    fillOpacity: 0.1,
    fillColor: 'rgba(125,139,143,1.0)',
  };
}
map.createPane('pane_contour_guinee_1');
map.getPane('pane_contour_guinee_1').style.zIndex = 401;
map.getPane('pane_contour_guinee_1').style['mix-blend-mode'] = 'normal';
var layer_contour_guinee_1 = new L.geoJson(json_contour_guinee_1, {
  attribution: '',
  dataVar: 'json_contour_guinee_1',
  layerName: 'layer_contour_guinee_1',
  pane: 'pane_contour_guinee_1',
  style: style_contour_guinee_1_0,
});
bounds_group.addLayer(layer_contour_guinee_1);
```

```javascript
pane: 'pane_region_naturelle_1',
    opacity: 1,
    color: 'rgba(35,35,35,1.0)',
    dashArray: '',
    lineCap: 'butt',
    lineJoin: 'miter',
    weight: 3.0,
    fill: true,
    fillOpacity: 0.1,
    fillColor: 'rgba(190,207,80,1.0)',
  };
}
map.createPane('pane_region_naturelle_1');
map.getPane('pane_region_naturelle_1').style.zIndex = 401;
map.getPane('pane_region_naturelle_1').style['mix-blend-mode'] = 'normal';
var layer_region_naturelle_1 = new L.geoJson(json_region_naturelle_1, {
  attribution: '',
  dataVar: 'json_region_naturelle_1',
  layerName: 'layer_region_naturelle_1',
  pane: 'pane_region_naturelle_1',
  style: style_region_naturelle_1_0,
}).bindPopup(function(layer) {
  return (layer.feature.properties.region_nat)
});
bounds_group.addLayer(layer_region_naturelle_1);
map.addLayer(layer_region_naturelle_1);
function style_region_administrative_1_0() {
  return {
    pane: 'pane_region_administrative_1',
```

```javascript
lineCap: 'butt',
    lineJoin: 'miter',
    weight: 3.0,
    fill: true,
    fillOpacity: 0.1,
    fillColor: 'rgba(225,89,137,1.0)',
  };
}
map.createPane('pane_region_administrative_1');
map.getPane('pane_region_administrative_1').style.zIndex = 401;
map.getPane('pane_region_administrative_1').style['mix-blend-mode'] = 'normal';
var layer_region_administrative_1 = new L.geoJson(json_region_administrative_1, {
  attribution: '',
  dataVar: 'json_region_administrative_1',
  layerName: 'layer_region_administrative_1',
  pane: 'pane_region_administrative_1',
  style: style_region_administrative_1_0,
}).bindPopup(function(layer) {
  return (layer.feature.properties.region_adm)
});
bounds_group.addLayer(layer_region_administrative_1);
map.addLayer(layer_region_administrative_1);
function style_limite_prefecture_1_0() {
  return {
    pane: 'pane_limite_prefecture_1',
    opacity: 1,
    color: 'rgba(35,35,35,1.0)',
    dashArray: '',
    lineCap: 'butt',
```

```javascript
fillOpacity: 0.1,
  fillColor: 'rgba(196,60,57,1.0)',
 };
}
map.createPane('pane_limite_prefecture_1');
map.getPane('pane_limite_prefecture_1').style.zIndex = 401;
map.getPane('pane_limite_prefecture_1').style['mix-blend-mode'] = 'normal';
var layer_limite_prefecture_1 = new L.geoJson(json_limite_prefecture_1, {
 attribution: '',
 dataVar: 'json_limite_prefecture_1',
 layerName: 'layer_limite_prefecture_1',
 pane: 'pane_limite_prefecture_1',
 style: style_limite_prefecture_1_0,
}).bindPopup(function(layer) {
 return (layer.feature.properties.nom_pref)
});
bounds_group.addLayer(layer_limite_prefecture_1);
map.addLayer(layer_limite_prefecture_1);
function style_limite_sousprefecture_1_0() {
 return {
  pane: 'pane_limite_sousprefecture_1',
  opacity: 1,
  color: 'rgba(35,35,35,1.0)',
  dashArray: '',
  lineCap: 'butt',
  lineJoin: 'miter',
  weight: 3.0,
  fill: true,
  fillOpacity: 0.1,
```

```javascript
map.createPane('pane_limite_sousprefecture_1');
map.getPane('pane_limite_sousprefecture_1').style.zIndex = 401;
map.getPane('pane_limite_sousprefecture_1').style['mix-blend-mode'] = 'normal';
var layer_limite_sousprefecture_1 = new L.geoJson(json_limite_sousprefecture_1, {
  attribution: '',
  dataVar: 'json_limite_sousprefecture_1',
  layerName: 'layer_limite_sousprefecture_1',
  pane: 'pane_limite_sousprefecture_1',
  style: style_limite_sousprefecture_1_0,
}).bindPopup(function(layer) {
  return (layer.feature.properties.nom_spref)
});
bounds_group.addLayer(layer_limite_sousprefecture_1);
map.addLayer(layer_limite_sousprefecture_1);

function pop_repartition_beneficiaires_projet_1(feature, layer) {
  layer.on({
    mouseout: function (e) {
      if (typeof layer.closePopup == 'function') {
        layer.closePopup();
      } else {
        layer.eachLayer(function (feature) {
          feature.closePopup();
        });
      }
    },
    mouseover: highlightFeature,
  });
  var popupContent = `<table>
```

```javascript
<td>${(feature.properties.femmes !== null ?
autolinker.link(feature.properties.femmes.toLocaleString()) : '')}</td>
            </tr>
            <tr>
                <th scope="row">jeunes</th>
                <td>${(feature.properties.jeunes !== null ?
autolinker.link(feature.properties.jeunes.toLocaleString()) : '')}</td>
            </tr>
            <tr>
                <th scope="row">total</th>
                <td>${(feature.properties.total !== null ?
autolinker.link(feature.properties.total.toLocaleString()) : '')}</td>
            </tr>
            <tr>
                <th scope="row">region_adm</th>
                <td>${(feature.properties.region_adm !== null ?
autolinker.link(feature.properties.region_adm.toLocaleString()) : '')}</td>
            </tr>
            <tr>
                <th scope="row">nom_utde</th>
                <td>${(feature.properties.nom_utde !== null ?
autolinker.link(feature.properties.nom_utde.toLocaleString()) : '')}</td>
            </tr>
        </table>`;
    layer.bindPopup(popupContent, {
     maxHeight: 400
    });
}
function style_repartition_beneficiaires_projet_1_0() {
 return {
  pane: 'pane_repartition_beneficiaires_projet_1',
```

```javascript
dashArray: '',
    lineCap: 'butt',
    lineJoin: 'miter',
    weight: 1.0,
    fill: true,
    fillOpacity: 0.1,
    fillColor: 'rgba(231,113,72,1.0)',
  };
}
map.createPane('pane_repartition_beneficiaires_projet_1');
map.getPane('pane_repartition_beneficiaires_projet_1').style.zIndex = 401;
map.getPane('pane_repartition_beneficiaires_projet_1').style['mix-blend-mode'] = 'normal';
var layer_repartition_beneficiaires_projet_1 = new
L.geoJson(json_repartition_beneficiaires_projet_1, {
  attribution: '',
  dataVar: 'json_repartition_beneficiaires_projet_1',
  layerName: 'layer_repartition_beneficiaires_projet_1',
  pane: 'pane_repartition_beneficiaires_projet_1',
  onEachFeature: pop_repartition_beneficiaires_projet_1,
  style: style_repartition_beneficiaires_projet_1_0,
});
bounds_group.addLayer(layer_repartition_beneficiaires_projet_1);
map.addLayer(layer_repartition_beneficiaires_projet_1);
function style_commune_cible_1_0() {
  return {
    pane: 'pane_commune_cible_1',
    opacity: 1,
    color: 'rgba(35,35,35,1.0)',
    dashArray: '',
    lineCap: 'butt',
```

```
weight: 3.0,
   fill: true,
   fillOpacity: 0.1,
   fillColor: 'rgba(133,182,111,1.0)',
 };
```

Printed by Books on Demand GmbH, Norderstedt / Germany